The Sheer Ecstasy

Of Being a Lunatic Farmer

To Lena —
Thanks for being
part of the healing,

Joel Salatin

i

The Sheer Ecstasy
Of Being a Lunatic Farmer

by Joel Salatin

Polyface Inc.

Swoope, Virginia

The Sheer Ecstasy of Being a Lunatic Farmer, First Edition
Copyright © 2010 by Joel Salatin

Editing and book design by Jeff and Carole Ishee

About the cover: Photo and concept by Rachel Salatin, graphics and design by Robin Leist.

The cover photo depicts Joel Salatin himself, the accessories he uses on the farm, and the sheer essence of "Lunatic Farming." If we must walk a mile in ones shoes to understand, then what better way to wear a farmer's gloves and hat to feel the hard work and passion of farming? "I grew up watching Dad use these tools and put his hat on every day, making them the expression of what this book is truly about," says Salatin. "Personally, I picture a smile under this rugged, bent, weathered hat; lying in the lush pasture content, just like the animals that walked the pasture just moments before."

Library of Congress Control Number: 2010909490

ISBN: 9780963810960

Other books
By

Joel Salatin

Pastured Poultry Profit$
Salad Bar Beef
You Can Farm
Family Friendly Farming
Holy Cows & Hog Heaven
Everything I Want to Do Is Illegal

All books available from:

Acres USA 1-800-355-5313
Amazon.com - www.amazon.com
Chelsea Green Publishing 1-800-311-2263

Or by special order from your local bookstore.

Table of Contents

Acknowledgements

Although my dad passed away in 1988, his spirit is very much alive here at Polyface Farm, and I first of all acknowledge his mentorship that so molded me into the person I am. Mom, still living here and very active, gave me the gift of theatrics and speech—the gregarious storyteller schmoozer. Thank you, Mom.

Who marries knowing exactly what they'll get? My beautiful wife, Teresa, could not have known the extent of my lunacy when we married 30 years ago. But she does now. And I think she still loves me.

Whenever I spend a winter writing a new book, my family and apprentices carry a heavier farm load so I can plunk away on the computer. Thank you Rachel, Daniel, and Sheri; and apprentices Ben Blasiman, Galen Menzel, and Matt Robertson. With a team like this, anybody can thrive being a lunatic.

As always, Jeff and Carole Ishee's desktop publishing magic took the raw pages and turned them into camera-ready copy. Robin Leist took my daughter Rachel's cover photo and combined her graphics arts skills for another wonderful book cover.

Mentors, customers, and the thousands of farmers who love being lunatics make my life *sheer ecstasy*. Thank you all.

Letter to Joel from Wendell Berry

Dear Joel,

Like me, you have a few opinions. I won't take a pledge to agree with every one of yours, and I would be disappointed in you if you agreed with every one of mine. We could have interesting discussions about the use of work horses, the proper role of government in agriculture, the issue of taxation, maybe a few other things. I would expect these discussions to improve my mind.

But the things we would need to discuss are remarkably few. I value this book highly, and I think everybody interested in the survival of authentic farming and farmers needs to read it. Here is the voice of experience, the voice of an actual farmer talking about farming. We have had too little of that.

I admire this book, first, for its unflagging, exuberant interest in details. The one legitimate reason to farm, as we know, is liking it—liking the work, even the smallest jobs, the smallest judgments and choices. It is the attention to detail that makes farming an art, and it pleases me that you recognize good farming as a fine art.

Second, I am grateful for your attention to the formal aspect of farming: the forming of the farm, the connecting of its various parts so that they sustain one another and become whole, which is another essential part of the artistry of farming.

Third, I admire and respect your insistence upon the connection of farming to its contexts of ecology on the one hand,

and of economy on the other. The products of farming, as you never let us forget, come ultimately from nature and, if the farm is to continue, they must go to paying customers. One of the two inescapable standards of farming is the health of nature, the other is the satisfaction (and the health) of consumers.

This is a book full of good sense, and when necessary it provides the indispensable wisdom: "Mark it down, if it smells bad or it's not beautiful, it's not good farming." Yes indeed.
Your friend,

Wendell

Written from Lanes Landing Farm, Port Royal, Kentucky, March 13, 2010.

Introduction

Every day I make thousands of beings happy. What a distinct privilege. Few people and few vocations present such an ecstatic opportunity. I love moving chickens, cows, and pigs because I know how happy it makes them.

One of my favorite chores is moving chicken shelters, both broilers and the eggmobiles. The unbridled delight these animals express through their demeanor and antics when offered a fresh salad bar is both obvious and palpable. You can feel the happiness in the flock.

The chickens can hardly dance fast enough. First an earthworm, then a cloverleaf. Oh, there is a cow pie, perfectly aged, with ten fat fly larvae waiting to be eaten. But on the way, a grasshopper jumps into view. Must get him now. Fly larvae can wait a few minutes. And a cricket on the way, suddenly still and quiet, hoping the little red hen won't notice. But she does, and she nabs the cricket on the way to the grasshopper and then scratches through the cow pie, pecking out fly larvae as fast as she can find them. Other chickens come over and greedily chow down on the tasty larvae. In seconds, the decimated cow pie has yielded its desserts and the hens are off for more from this fresh morning salad bar.

Every afternoon, when I move the cow herd into their new salad bar paddock, the contented chewing, grass ripping, and cavorting of the calves epitomizes blissful joy. Earthworms wait expectantly for the cow dung and the newly released root organic matter from self-pruned plants. Every day the seen and unseen life on the farm responds gratefully and exuberantly to my care. What sheer ecstasy to be able to make this many beings happy.

Unfortunately, our Greco-Roman western linear reductionist systematized fragmented disconnected parts-oriented individualized culture does not make these critters happy. And it considers anyone who reaches for such a goal to be a lunatic. We're supposed to be interested in growing it faster, fatter, bigger, cheaper. Nothing else matters. And all these beings are just inanimate piles of protoplasmic structure to be manipulated however cleverly hubris can imagine to manipulate them. Yes, that's the American way. Truly patriotic.

A few weeks ago I ordered a tractor-trailer load of sawdust from a nearby sawmill only to be informed that they had subcontracted out that service. The receptionist happily gave me the number of the fellow who made those deliveries and I called him immediately.

"Hi, this is Joel Salatin. I need a load of sawdust for barn bedding."

"Great. When do you want it?"

"Monday would be fine. Does that work for you?"

"Sure. I'll bring it about 9 a.m. on Monday. Thank you very much. See you then."

"Bye."

All was well. Two hours later, I answered the phone. Same guy, different tone.

"I didn't realize who you were but your name sounded familiar so I checked into it and found out you were THAT guy."

"Oh, what's the problem?"

"I wouldn't bring you sawdust for a million dollars. You abuse your cows because you don't give them hormones, so they can't grow as fast as they should. You don't feed your cows grain, so they are stunted. You abuse your chickens by not having them in an environmentally controlled house. You're going to bring back hog cholera and disease by letting your pigs run outside...."

By this time, the phone was melting in my hand as he went on with the tirade. I said okay and hung up. I ended up having to go outside our local area to get the sawdust—to somebody who was not familiar at all with Polyface Farm.

Amazingly, the farms that dump on chemicals, dope their animals, confine their animals in factory farms without fresh air, sunshine, and salad bar are now considered normal and I'm the lunatic. As the industrial food system grows, I realize more and more how different my paradigm is, on many levels. We are not simply a preference apart. We are not just different nuances of the same thing. We are on different planets. In fact, we are on a collision course. We are at war.

I believe some things are right and some things are wrong. I think some ideas are right and some are wrong. I think a dark side does exist. And I don't want to be a part of it. I can't imagine working for outfits trying to extract the porcine stress gene so we can disrespect pigs even more but at least they won't be stressed about it.

I can't imagine working for Monsanto, a company that sues farmers for patent infringement when Monsanto's genetically engineered pollen trespasses next door and impregnates the neighbor farmer's crop. That's not about preferences. That's evil.

But rather than being downtrodden and depressed, I'd rather enjoy encouraging, developing, and living in a more righteous system. What I've learned is that what Monsanto and company (which includes the United States Department of Agriculture, the Food Safety and Inspection Service, the Food and Drug Administration, other globalists and most politicians) considers lunacy is actually sheer ecstasy. To watch our animals exuberantly dance into their salad bar rather than slink back sullenly against cellmates in a fecal factory concentration camp farm. To enjoy customers and visitors interacting with the farm, building relationships and memories that will carry integrity to the dinner plate. That sure beats NO TRESPASSING signs and BIOSECURITY warnings. It sure beats having to don a hazardous material suit and walk through sheep dip just to visit the animals.

Trust me; the industrial food system is not looking at the local integrity food movement wishing they could be like us. The industrial food system feels threatened by everything we stand for. They truly believe, deep down, that they must save the world from people like us: people who believe we should use compost, biologicals, and pasture-based livestock production models.

Industrial foodists believe farmers like me are barbarians and Luddites. They really believe that my ideas threaten civilization's health and progress. What I consider liberating, they see as abusive. Indeed, to them, I'm a lunatic.

But in my lunacy, I enjoy increasing carbon sequestration where they deal with tighter soils and decreasing organic matter. I'm less dependent on petroleum while they become more dependent by the day. I have fewer pathogens while they wrestle with bovine spongiform encephalopathy, E. coli, salmonella, campylobacter, and listeria. And while my customers enjoy vibrant health, theirs develop Type II diabetes and obesity. While they put their faith in pharmaceuticals, I dance with functioning immune systems. While they frown, I laugh.

Indeed, I'm having the time of my life. Earthworms dance in my fields. Pigs fully express their pigness. Eggmobiles sanitize the fields. Enthusiastic young people work and learn. Cows practice their mob stocking herbivorous solar conversion lignified carbon sequestration fertilization. It's ballet in the pasture. Aesthetically and aromatically sensually romantic. It's sheer ecstasy.

Folks, the worldview of the locally-based, community-imbedded, environmentally-enhancive, nutrient-dense farmer is as different from the worldview of globalist, mega-corporate, industrialized food elitists as East is from West. As Native Americans from Europeans. As libertarians from liberals. As Beauchamp from Pasteur. As Incas from Conquistadors. As righteousness from evil.

My goal in this book is to encourage the righteous side and to have fun doing it. Let's embrace being lunatics. Most great prophets were considered lunatics in their day. His contemporaries thought Jesus had a devil. Our side can either respond with hand wringing and despondency, or we can have fun enjoying our answers. Our truth. We can enjoy the sheer ecstasy of being lunatic farmers, either by proxy or for real. Welcome aboard.

As the industrial food system lashes back with innuendo and pseudo-science against the local ecologically based food system, I think it behooves all of us to examine the differences between these two camps. People wonder how I can be such a happy farmer. The stereotypical complaining, unhappy farmer unfortunately is true much of the time. I hope this book will help put in clear detail the depth and breadth of the difference between the chemical/industrial/global approach and the local/biological/ecological approach.

I hope that you will catch the vision. If the lunatic fringe is where the truth is, I hope you will aspire to be a lunatic too. Let's enjoy it, embrace it, and have fun with it. Indeed, let's enjoy the sheer ecstasy of being lunatic farmers. Or at least patronizing one. Now let's have fun.

JOEL SALATIN

Sept. 1, 2010

Nurture the Earth

Chapter 1

Growing Soil

I live in Virginia's Shenandoah Valley, the proverbial "breadbasket of the Confederacy," where the ravages of soil disrespect are everywhere apparent. Because this 16,000-square mile valley was the tall grass prairie of the East, it naturally became the first intensive grain production region. By the 1870s the eroded hillsides gave way to protruding rock outcroppings and the heart of grain production moved to the Midwest.

When our family arrived on this farm in 1961, deep gullies grooved all the slopes. Large, shallow craters in the fields had no topsoil whatsoever; these large galls were solid shale rock. Some were nearly 100 ft. in diameter. Locals estimate that most hillsides in the valley lost between 3 and 8 ft. of soil between initial prairie breakup and today. If you think about it, losing 1/10 of an inch of soil per year is an inch a decade. From 1740 until 2010 is 27 decades, which at a minimum inch per decade is 27 inches at least. And under the severe assault of translocated, European-style tillage farming in those early decades, certainly a few years saw 1/4 inch of soil loss.

When Dad began developing portable electric fencing systems in the early 1960s these galls and thin soils were not deep

2

enough to hold up an electric fence stake. These were not fence posts. They were not even survey stakes. They were just little 3/8 inch rebar 4 ft. long to hold up a thread of wire. Not enough soil.

He poured cement into old car tires and pushed two pieces of half inch pipe into the cement before it hardened: one vertical and one slightly off vertical. This connivance offered a stanchion and stake-holding sleeve to hold up the stakes where no soil existed. When we installed or moved electric fence, Dad would heave a dozen of these concrete tire bases onto a tractor-mounted carrier and my brother and I would heft them off as we drove through the field. Then Dad would come along and insert the stakes into the pipe sleeves. Once the stakes were in place, he would roll out the wire, placing it in insulators on the stakes.

Talk about hard scrabble farming! Over the years, by rotating the cattle, applying compost, and running poultry, these shale galls began healing. They heal just like a wound on your skin. Around the outer edge, a little berm of scab develops. Eventually the hemorrhaging (erosion) stops when the scab stretches all the way across the wound. Then new skin develops at the outer edge of the scab. After some time, the final piece of scab sloughs off and new skin glistens underneath.

These galls went through the same procedure. Every year new soil rolled into the edge of the scalloped shale rock. These galls were saucer-like shallow depressions of solid rock pocking the field. Gradually a few weeds found a toehold in the seams between the shale layers. The dip from soil into the depressions created a ticklish navigational hurdle for haying equipment.

Each year, up through the 1970s and 1980s, the soil marched another foot toward the center, reclaiming bit by bit the productive capacity lost through decades of irreverent abuse. By the year 2000, after 40 years of our family's stewardship, all of these galls had been reduced to a couple of feet in diameter. Today, they are completely enclosed and we can hand-push an electric fence stake 8 inches into the ground on every inch of those areas. My grandchildren will never know, much less remember, where those galls were. They do not have to build electric fence with concrete tire stanchions. Grass waves lush and green on every square foot of the field, not just the areas between rock craters.

What a tragedy that my grandchildren will not know the joys of broken hay mowers, knocking teeth off the hay rake, and bouncing off of a load of hay across this pock-marked landscape. All they will know are lush fields, black soil, and profit. And I will enjoy sheer ecstasy, watching them reap bountiful solar-produced biomass from these productive fields.

Now to the gullies. These, of course, developed long before municipal landfills. Where did farmers put trash? In the gullies. Each farm ran its own mini-landfill operation. Bottles, boots, mattresses, bedsprings, old equipment—you name it; it's in the gullies. As money became available, we hired an excavator to build ponds in the low ground and carry that soil and silt up the hillside to the gullies. In the low lying areas, old fence posts that used to stick out of the ground 5 feet are now completely submerged by years of accumulated silt.

How did they get covered up with soil? That's the soil that washed off the hillsides, gullies being the most obvious monument to the shift in soil location. Removing the accumulated decades of debris from the gullies was too large a task, but we at least covered it up and recreated the gentle, undulating fields that were here when bison covered the landscape.

Digging the silt out of the low areas created ponds for water catchment. The excavation, then, accomplished two things with one task: ponds for additional water storage and gully filling.

Other gullied slopes were too steep or numerous to heal this way, so we planted trees on them. Today, the gullies are still there like corrugated roofing, but at least they are covered in trees and a layer of mulch. One gully was 16 feet deep. Imagine how much soil washed down to the Chesapeake Bay to create a 16 foot deep by 150 foot long gully. The volume is equivalent to a house. And this is just one gully of many on one farm of many in one valley of many in one county of hundreds within the waterhsed.

All my life I've worked on and around these eroded areas and wondered: "What were they thinking?" Indeed, those first Europeans, those Scotch-Irish Presbyterians who settled this hamlet—"What on earth *were* they thinking?" And then as if the initial insult weren't bad enough, what were their children thinking? And when the gullies started, what were those farmers thinking? And as the gullies grew bigger and they finally quit plowing the fields because they couldn't navigate the 10 foot deep gullies, up and down, up and down, what were they *thinking*?

Why did it take a family from outside the area to be appalled at this devastation? Why was the previous owner still plowing in these eroded fields, still grazing these gullied hillsides? When Dad and Mom bought the farm, these hillsides literally contained not a shred of vegetation for entire acres. Bare clay, still eroding. Muddy water still gushing down the streams. Still pushing out into the Chesapeake Bay. Why were cattle still marauding these gullies, still uprooting the stray weed, still promoting a denuding, actively eroding landscape?

Was it because these farmers had no money? Of course not. These successful farmers built fine southern homes and enormous barns. They drove fine trucks and fine cars. They had sofas, kitchens, toys and plenty of clothes. Even a Sunday suit. The three successive owners from 1948 to 1961, when our family came, were all people of means. They cavorted with the local blue hairs, shared wheeling and dealing parties, and put money in the offering plate at church.

But let's go back further. How about those first settlers? The first Europeans to lay eyes on the Shenandoah Valley described a silvo-pastoral landscape - a veritable sea of grass so tall that the blades could be tied in a knot above the horse's saddle. This savannah with widely spaced trees amid thick grasses had been created and maintained for centuries by large herbivore herds (buffalo, elk, deer) systematically chased by predation (wolves, fire, or Native Americans). This time-sensitive landscape management, using disturbance and subsequent rest periods, created a perfectly suited soil building landscape.

The periodic disturbances of grazing or fire massaged the landscape into successive regeneration. All biological communities need periodic disturbances to freshen them up, to succeed to climax ecosystems. Call it ecological exercise, if you will. A lethargic couch potato ecosystem is stagnant. Most designated wilderness areas are stagnant. Eventually, the decaying biomass accumulates enough to support a fire that even local fire departments can't extinguish. The conflagration refreshes, renews, and exercises the ecology into regenerative production.

When the Europeans arrived, they did not ask the natives how to do this. The newcomers came from a temperate climate with gentle rainfall: "one misty, moisty morning, when foggy was the weather..." The wide temperature fluctuations, violent summer thunderstorms, and the fragile clay-based soils were not conducive to the kind of broad acre, arable farming practiced in the British Isles and continental Europe. But without asking, without taking stock of the new realities, these newcomers plowed up every square yard they could find.

Early on—very early on—farmers began lamenting soil fertility loss. George Washington, Thomas Jefferson, John Taylor from Caroline and hundreds of other farmers during the colonial period wrote about soil loss. But they were planters. And the planting class was here to wrest from the perennial-herbivore-silvo-pasture a new domesticated landscape without historical reference to this new place. I have read their writings and find it amazing that in all their seeking, they did not seek counsel from the natives. After all, they thought the natives were just barbarians who didn't powder their wigs or abide by parliamentary procedure.

The natives used talking sticks and lived in tribes. What would these barbarians know, after all? What kind of wealth could be created without exporting tobacco? As the newcomers divided the land into private holdings, the disturbances through tillage became not only more severe but the rest periods between disturbances became too short. Failing to understand the life and magic in soil, they viewed it like an inanimate object to be exploited.

True stewards of the land understand that more living organisms are in a double-handful of healthy soil than there are people on the face of the earth. With modern electron microscopes, the soil community is being brought out of its secrecy by researchers around the world. Live video of this soil community, or what Elaine Ingham calls "the soil food web," is more profound and other-worldly than the most far-fetched science-fiction movies imaginable. Multi-legged critters with antennae stab the side of other complex micro-organisms.

There in the space of a pinhead thousands of teaming micro-organisms vie for ascendancy, eating and being eaten. Stabbing and jabbing. Sucking body fluids and scissoring off the heads of foes. And mating and having babies and going to kindergarten and encountering the schoolyard bully. And suing and divorcing and driving around in little cars...okay, let's not get carried away. The point is that the commotion going on in the soil apes anything we can imagine on the surface.

Wars, conferences, Dilbert cubicles and family Christmas drama cannot compare to the soil's theater. It's beyond anything you can imagine. And it's been going on since Creation. And unless disrespectful soil exploitation destroys it, it will continue apace until the end of time. And this precious resource of mineral, decaying biomass, gasses, water, and critters is the only protective veil between humanity and starvation. Compared to the earth's mass, it's equivalent to sheer lingerie. Utterly dependent on this most precious resource, we arrogantly dismiss it as dirt. And people think it's icky.

Soil is in fact marvelous beyond description, and the thought of having less of it on our farm than last year would make me prostrate in repentance, remorseful beyond description. I can't imagine living with less of it. I can't imagine destroying it. I can't imagine abusing it.

So what *were* these predecessors thinking? I've asked many people that question, and my conclusion is that they simply weren't. They weren't *thinking*. They were going about their routines, completely blind to their most precious asset. Oblivious to their most important responsibility. Ignorant about how to grow it. As cubic yard after cubic yard washed down the river, they went shopping. They exchanged Christmas gifts. They listened to sermons. They went about their daily lives and never thought about it. It was the most obvious and cataclysmic dispossession they could imagine, but they didn't think about it. Business as usual.

If someone had taken their horse, they would have moved heaven and earth to chase down the culprit and see him brought to justice. If someone had stolen their money, they would have held the perpetrator at gunpoint until the sheriff arrived. If they had a tear in their trousers they would have spent an afternoon fixing it. If they had a hole in their socks, they would have darned it. But every day the accumulated wealth of centuries was cascading off their farms, and they did not stop it.

7

But it gets worse. Today, nearly 3 centuries later, we know about the magical, living, beautiful community in soils. We know about organic matter. We know about sheet erosion, dust storms, rill erosion and a host of soil-related information. We are wealthy beyond any previous civilization's wildest dreams: flat screen TVs, iPhones, Blackberries, Lexus and Tom-Toms. We are rolling in luxury.

Yet for all of this, we are losing soil faster than at any time in history. Thousands of farmers right here in the Shenandoah Valley continue to plow hillsides—the more gentle ones now since the steeper ones have been abandoned to forest reclamation. They apply chemical fertilizer that burns out organic matter, which is the secret to all soil life. The rivers still run muddy even after a half inch thunderstorm. Gullies continue to grow and rock outcroppings enlarge as the soil around them loses its battle with gravity. Overgrazed pastures with terraced, hillside cow paths outnumber rotationally-grazed pastures a hundred to one.

Here on our farm, we realized as soon as we arrived that these fragile slopes could not be plowed. That perennials built the soil in the first place and perennials will heal it. That animals in all their dimensions - from dung to hoof aeration and scratching - massaged the vegetation into proliferation. And that decaying biomass is what builds soil. Long before petroleum-based fertilizer; long before John Deere invented the moldboard plow; long before Cyrus McCormick invented the reaper; long before diesel engines and hybrid seed, solar-built plant material was feeding and creating soil through the decay process.

While other farmers built their farms around annual plants, we built ours around perennials. The effort in labor and energy required to till, plant, fertilize, weed, harvest, and store annuals accounts for the lion's share of expense in American agriculture. Moving the billions of tons of soil in tillage. Or spraying herbicide to kill vegetation in order to plant no-till. The fuel, machinery, infrastructure, and labor required to grow annuals is astronomical compared to perennials.

Shortly after moving onto this newly-acquired farm, Dad asked several credentialed expert farm advisors to come and tell him how to make a living on this farm. Universally, public and private alike, advised: plant corn, build silos, graze the forest.

8

Dad was smart enough to spurn every one of those expert opinions as violating economics or ecology, or both. Viewing nature as a template, and appreciating the herbivore-perennial-disturbance-rest relationship, he set about to duplicate those principles instead.

Only a lunatic would embark on such a contrarian course. Neighbors laughed us to scorn. When Dad said that some day we would have 100 cows on this farm, you'd have thought he was promising to fly a broomstick to the moon. I've often wondered what kind of conversations must have ensued in those farmers' homes after these encounters. Do you want to know a secret? Not only has our farm reached those numbers; we've exceeded them. Oh, the sheer ecstasy of being a lunatic farmer. While others struggle with annuals, we just move cows around. While they spray weeds with herbicides, we let the cows eat the weeds.

While we move some electric fence every day, they plant corn, fill silos (bankruptcy tubes), and worry about commodity grain prices. We don't own a plow, a disk, a planter, a corn chopper. I was giving a talk to a Ruritan Club many years ago and after I'd finished, an old farmer on the front row, arms crossed resolutely across his chest, verbally assaulted me: "Let me get this right. You don't have a plow. You don't fill a silo. You don't combine wheat. Well, sonny, you don't do any farming, then, do you?" Annuals are imbedded in our agrarian roots to such an extent that to even consider a perennial-based agriculture is an assault against farming.

But the perennial based system is what built these soils. And it's what built the soils of the Midwest that are just 100 years behind being mined out like the ones here in the Shenandoah Valley. This is one reason why Wes Jackson at the Land Institute in Kansas has devoted his life to experimenting with perennial grain crops. Perennials typically put more energy into biomass than seed. Annuals put all their energy into seed (the grain). If we can breed plant perennials that could displace annuals for seed production, it would be a breakthrough indeed.

Long before Justus von Leibig told the world that plants were only nitrogen, potassium, and phosphorus, plants were enjoying iron, boron, selenium, magnesium, calcium and all the other minerals. Just because nobody at the time saw the intricacies didn't mean they weren't real. And just because the world scoffed at J. I. Rodale and the pioneers of the organic movement who disagreed with such simplistic analysis didn't mean the credentialed soil scientists were right. They were, in fact, wrong.

And finally in the 1970s and 1980s simplistic NPK fertilization was universally discredited. And yet, that NPK mentality still dominates the entire farming system. Even with all we know, that's the foundation of fertility programs, even though it's well known to be inadequate and soil depriving. And to add insult to injury, today's agronomists and plant geneticists have become sophisticated enough to quit thinking about soil and talk about plant food. The soil is just an inert substance to hold up the plant. The whole goal is to hook up a plant food intravenous tube, so to speak, and mainline petroleum-based concocted chemicals right into the plant's arteries, sans roots and soil. Fertilizer is out; plant food is in.

Dad saw the fallacy of this thinking before Rachel Carson wrote *Silent Spring*. He saw the chemical fertilizer approach as a grandiose drug trip, always requiring heavier doses to get the same kick. Pesticides, herbicides, and fungicides created a treadmill requiring a more toxic dose to achieve the same high. As a result, we began doing everything possible on our farm to recycle manure. And we began composting. Big time.

While other farmers scurried around trying to buy the latest petroleum concoction to stimulate soil fertility, we bought a chipper and began turning tree branches into fertilizer. The forest became our fertilizer factory. We didn't need the Arabs. We had a whole solar collector right outside the back door. And those trees brought up deep minerals from the soil so that the chips had a natural balance and complexity of minerals to feed the soil community's varied and voracious appetite.

While others patronized train car loads of material from the Middle East, we began diverting leaves, sawdust, yard wastes, Christmas tree chips and anything else decomposable from the landfill to our composting operations. We even paid right-of-way maintenance crews to dump their loads of tree chips here at the farm. We are fiends for organic material. Dad brought home truck loads of corn cobs from the grain elevator and spread them on the land. This was before combines were in wide use.

My ticket away from off-farm employment September 24, 1982 was a direct result of looking for organic material. A black walnut buying company in Missouri expanded to Virginia in 1981 and set up a buying station in Staunton about 12 miles away from our farm. I was working as a journalist at the local daily newspaper at the time and did an article about the new set up. It was being operated on Saturdays by two FFA boys from a local high school.

Located on the back parking lot of the local Southern States dealer, the boys—and their dads—were struggling with the walnut hulls. Each walnut has a soft, nitrogenous outer hull that a machine separated from the hard nuts. The Missouri company hauled the hard nuts from Virginia to their processing facility and cracked out the meats. I knew that grass always grows great around a walnut tree, so I asked the fellows if we could get some of the hulls. "Take all you want, for free," they said.

The next Saturday I sent Dad down there with the dump truck and he brought home a load. We didn't have a front end loader at the time, so I hand shoveled the 6 tons into a manure spreader and applied them. We brought home as many as we could that fall. The next spring, the hull-fertilized pasture grew hay 7 feet tall. The Southern States dealer had struggled with the traffic congestion created by the once-a-day operation and decided a 6-day-a-week schedule was necessary. I asked to do it and they agreed. The next fall, in 1982, I cleaned out my newspaper desk on Friday and began hulling walnuts on Monday.

And that's how I started fulltime farming—getting paid to apply carbon. As this organic matter and composting system fed the soil, it responded magnificently. So much so that the fields of my youth, which grew such poor grass that we could scarcely cobble together 10 small square bales (40 lbs. apiece) of hay per acre, now yield 150 bales. The farm that 50 years ago would scarcely support 10 cows now generously supports 100-plus.

Our local equipment dealer recently encouraged us to buy a new wheel rake for making hay. It's the new rage and everyone is buying them. We asked for a demonstration. The salesman brought it out and headed down the field. Our hay was so thick that the brand new state-of-the-art machine—the one everyone else thought was the greatest thing since sliced bread—couldn't rake

11

our hay. It just gummed up, clogged up, and no amount of adjusting would fix it. And that was after we'd already grazed the field twice early that spring. The salesman just shook his head and muttered: "I've never seen anything like this." But we're lunatics, see. We've been scrounging organic matter while everyone else was spraying and applying chemical fertilizer.

We had a neighbor come down to mow the hay with a haybine because we couldn't get through it with our cycle mower. A cycle mower has serrated knives that reciprocate between plates to shear off the grass stems. He couldn't mow it any better than we could because the earthworm castings were so high they clogged up the knives. Finally we called a neighbor with a discbine. This is a machine with saucer-type disks with knives attached to the outside edge of these spinning disks. Disc mowers were developed to handle ant hills and thicker forage.

He was an old time farmer and had mowed lots of fields in his life. Located just a mile away from us, he was well familiar with our lunatic methods. No doubt had cracked a few jokes about those lunatic Salatins over the hill. When he got done mowing—and his machine performed beautifully—he just shook his head and said: "I never knowed nobody could grow hay with mulch." That's what he called our compost—mulch. Isn't that hilarious? I laughed so hard I almost fell off the tractor. Compost is a foreign word to most farmers.

Actually, the more I thought about it, he wasn't that far off. Landscapers understand the value of mulch around flowers and trees. Gardeners understand the value of mulch around vegetables and fruit. Why can't farmers realize the value of mulch? Only lunatics like me see any value in it.

Since chemical fertilizer burns out the soil organic matter, other farmers struggle with tilth, water retention, and basic soil nutrients. The soil gets harder and harder every year as the chemicals burn out the organic matter, which gives the soil its sponginess. One pound of organic matter holds 4 pounds of water. The best drought protection any farmer can acquire is more soil organic matter. Most farmers, however, would rather scurry around trying to buy drought protection through insurance programs. Goodness, if they'd just increase their organic matter, half the things they battle would just go away.

It's as if the whole notion of growing soil is something only lunatics would think about. But why not grow soil? Does anything make more sense than growing soil? Isn't that more important than tractors, trucks, silos, barns, county fairs and country music? Of course it is. And yet to the lion's share of American farmers, the very notion of growing soil is just plain silly.

Everything we do at our farm is geared to growing soil. We put the livestock mineral box on thin patches to get the benefit of the concentrated dunging in that area. Most farmers around here have a stationary mineral feeder that denudes the soil and ensures that all dung will run off in the next rain. And generally it's located in an easily accessible low area instead of up on a hill where the dung would do more good. My goodness, I won't even pee on a neighbor's land. If I'm building a boundary fence, for example, I'll make sure I step over to *my* side to add my fertility. Is that greedy? No, it's just being passionate about growing soil.

We rent several farms and when we begin managing new farms, they always have denuded hay or silage feeding areas. Feeding areas, of course, concentrate dung because the animals spend more time there eating. These feeding areas need to be moved around in order to spread the manure. Instead, most farmers just put feeders in a handy spot and leave them there year after year. That deprives the adjoining land of fertility and over applies manure on that particular area. That over-application results in the fertility either leaching into the groundwater, running off into the stream, or vaporizing off into the atmosphere—or all three.

Several years ago I attended a class taught by a Virginia Tech agriculture economics professor about whether or not to sell fall calves or overwinter them and sell them in the spring. He had a nice chart comparing feed costs and sale prices. The idea was that if hay was priced high and calves were selling low, it would be better to sell the calves and sell the hay. If the hay is low and calves are high, it may be better to feed the hay and sell the calves on the high spring market. He went through a whole pile of scenarios and benchmarks.

When he opened the room up for Q&A, I raised my hand and asked if he had considered the fertility value of the dung by retaining calf ownership and feeding the hay through the winter.

"Oh, I never thought of that," he said. Of course not, because only lunatics think about growing soil, which means closing the fertility leaks on farms. But if your paradigm is that fertility comes from somewhere else, and it's something you buy and don't create on the farm, then retained manure value never even enters the equation.

Years ago I used to do some farm consulting. The first thing I did on every single farm was a manure audit. Typically, the farm would be generating 5 tons per acre of manure, which is enough to grow soil. But then on the expense ledger, they were spending $15,000 on fertilizer. Why? Because it was going down the creek, in the groundwater, off in the air stinking up the neighborhood.

The joy of knowing that every day our farm is growing soil is beyond description. It is getting better and better, not worse and worse. We're not wearing out the farm. We're massaging the soil to new levels of vigor and health. That's the sheer ecstasy of being a lunatic farmer.

TAKEAWAY POINTS:

1. Herbivores and perennials are the most efficacious way to build soil.

2. Tillage and annuals account for the lion's share of erosion. .

3. Depleted soil can be rebuilt and regenerated.

4. Carbon is the key to soil health.

Chapter 2

Grass Farmer

When the wolves chased buffalo across most of North America, or native Americans started fires, or the lions licked their chops around cape buffalo in Botswana, they were grass farming. It's the oldest carbon sequestration and earth breathing choreography we know. All nomadic tribes, from ancient Hebrews like Abraham, Isaac, and Jacob to Ghenghis Khan's Huns from outer Mongolia to the gauchos of Argentina's pampas to the Swahilis in Africa—are ultimately shepherds.

Herding livestock is the domestic equivalent of the predator-prey relationship that climaxes every environmental program. The show starts harmless enough but builds and builds until that final violent kill. Dust flies. The herd scatters, tails high in the air. The lion or leopard in that tremendous burst of agility and raw strength crunches neck bones on the gazelle and it goes limp. The big cat turns her face toward the camera, blood oozing from her lower jaw, whiskers wet with dew and body fluids, ribcage heaving as she recuperates from the exertion. That, dear friends, is grass farming.

It's as ancient as history. Far more primal than grain farming. Some Biblical scholars look at the curse of Adam and

Eve in Eden and point out that the woman's curse was the pain of child bearing and the man's curse was tilling the ground— indicating that prior to that time, grain had not been growing.

Some of the most beautiful word pictures in the iconic *Little House on the Prairie* books are centered around descriptions of the prairie. Undulating 8 foot grasses stretching to the horizon. Sitting in a covered wagon, trundling along at an oxen pace what a sight that must have been. Several years ago I had the privilege of speaking at the University of Nebraska, Lincoln. The college there maintains an acre of native prairie. I walked over into it and it had a profound affect on me. A veritable religious experience. If ever I walked on holy ground, that was it. Similar to the redwood forests in California, but more powerful for me.

Surrounded by 8 foot tall grass, with stems as thick as your thumb, I was overwhelmed with a sense of smallness. Insignificance. Just a few feet into the field I was completely lost to the outside world. Towering over me these blades and seed heads of grass blocked out every distant reference. When Laura Ingalls Wilder recalled in her books how afraid Pa and Ma were that the girls would be lost in this vast sea of fronds, I could easily imagine that parental terror. Anything could creep up on you. Were it not for the paths through this little field, about the size of a football field, you could literally go 20 yards in and not find your way out.

As I thrilled to the rustling of that mighty grass sea, I closed my eyes and imagined a herd of 3 million bison, undulating in waves, chomping, stomping, butting, cavorting, jockeying, eating, eating, eating, pooping, pooping, pooping. It reminded me that for all of our plowing, fertilizing, machinery, and farm energy, modern American farming still does not produce the amount of meat that was produced here a few centuries ago. Most of that meat was consumed by wolves or destroyed in incredibly inefficient hunting techniques by native peoples.

Native Americans would light fires to stampede a herd off a cliff. Imagine how high a cliff would have to be for a buffalo to actually be killed tumbling off it. These are tough critters, and they can take a lot of rough and tumble. Lots more than professional wrestlers. So many would die that the cliff edge would fill up with dead bison until the last of the herd just ran out over the accumulated dead carcasses. Corn and petroleum fertilizer can never, ever, compete with the production of the native prairie.

Imagine my beloved Shenandoah Valley covered with that kind of grass. The vast Midwest. Even Colorado and California. One of our apprentices 20 years ago said that his grandmother told him stories about traveling up through the state of Sonora, Mexico when she was a little girl. From a monied family, she attended boarding school in British Columbia in the early part of the 20th century. All up through Arizona she said you couldn't see the landscape due to the tall grass growing up along the edge of the road. Today, when you enter that land, from Douglas, Arizona, down through Agua Prieta and through Sonora, grass doesn't exist. It's barren, eroding, desertifying. Raw, open soil baking in the sun.

The grass that built the West, that created the wealth that built ranches and legends—it's gone. This most precious green blanket, meant to clothe the soil and protect the soil community from baking sun and torrential rain—it's gone. The millions of bison, antelope, elk, prairie chickens, and pheasants that mobbed across and received sustenance from this grass treasure—they're gone. Water that bubbled up from springs and gently-sloped hoof-excavated and manicured stream banks meandering through the grassy meadows—they're gone. Instead, stark arroyos cut deep into baking earth speak of a sick and dying hydrologic cycle. No sponge to accept thunderstorms.

The wealth of any ecosystem is its perennials. The primal herbivore-predator-disturbance-rest dance is literally the breath and pulse of the earth. Grasses recycle oxygen far more efficiently than trees. The turnover is faster. Grass reaches out and turns solar energy into carbon. Tillage hyper-aerates the soil, burning out carbon. But because a plant creates bilateral symmetry at the soil horizon, it sloughs off root mass when the top gets chopped off.

This voluntary pruning preserves energy in the crown, or core, of the plant. When you buy a 2-year-old apple tree from a nursery, it looks like a stick. The transplant shock is minimized by pruning off appendages and concentrating the plant life energy in the core. If you jump into freezing water, your body voluntarily shuts down extremeties to preserve core function: heart, lungs, liver, kidneys. Not having to put precious energy into appendages shepherds core functions.

This is exactly what grass does when it's mowed mechanically or by grazers, or when it's burned. What's below the soil mirrors what's above the soil. That's why you don't want to mow your lawn too short. The shorter you maintain the grass, the more prone it is to drought because the roots don't go anywhere. Leaving some height and letting it grow longer allows the roots to go deeper and find moisture.

The ancient art of grass farming, at its most fundamental level, massages this energy flow and biomass accumulation. Unfortunately humans have massaged incorrectly or with pretty rough hands all too often. I'm well aware that more often than not human understanding of this ancient carbon-accumulating dance has either been misunderstood, spurned, or adulterated. Overgrazing and carbon depletion is, unfortunately, far more normal than carbon accumulation.

But that is no reason to demonize cows any more than it's appropriate to demonize priests because a couple of them are pedophiles; or teachers because a few of them are inept. The reason I belabor the ancient primal predator-prey relationships is because that is the foundation for the heartbeat of the earth's carbon, hydrologic, and oxygen cycle. The grass, as it stretches and lengthens, grows faster and faster before finally slowing down at seedhead and senescence. Encouraging it to go through its juvenile growth spurt is the key to accumulating carbon both above and below the soil. It has to create a carbon bank.

As this rapid and fully expressed phenotypical growth capitalizes on solar energy and accumulates carbon, transpiration kicks into high gear pumping out oxygen. The grass breathes in carbon dioxide and exhales oxygen. When the herbivore eats off the top of the plant, the roots prune back and pulse organic matter into the soil. This soil accumulation of carbon, ultimately, is any ecosystem's health plan. And a culture or civilization can never be any healthier than its soil carbon health plan.

So how do we mimic this ancient earth cardio-pulmonary function? First, it requires animals. Lots and lots of different kinds of animals. It requires aggressive movement and grazing. It requires times of rest to let the grasses accumulate carbon and energy.

Why this ancient relationship is not more revered never ceases to amaze me. Men swagger around calling themselves "cattlemen" but abuse their grass like a rapist. And abuse their cattle with concrete fecal feedlots without any regards to rumen function. Vegetable growers plow thousands of acres, planting monocrops of annuals in a never-ending tillage routine that totally annihilates soil carbon wealth. Why? Why are we so enamored of things that destroy carbon and disrespect the animals under our care?

Grass. Lowly grass. It just gets no respect. And yet it is the lifeblood of the planet. If you revere grass like I do, the culture calls you a lunatic. Who could ever get excited about grass? It's just grass. It stains our soccer pants. We drag it in the house on our shoes after mowing the lawn. If it's there, great. If it isn't, big deal. Unfortunately, the great grasslands of the world have not only shrunk to veritable obscurity, but they have never attracted resorts and tourists.

Recreating this carbon cycle through the high metabolism of grass is the obsession of lunatics like me who affectionately call ourselves grass farmers. We wear the mantra proudly. We see ourselves as the earth's true physicians, trying to restart the most basic cycles that maintain equilibrium between land, air, and water. Western cultures have prestigious awards for corn growing and cattle breeding and strawberry cultivation. Even the American Forest Council has an American Tree Farm System to laud the efforts of landowners who steward their trees.

But grass? Yes, there is an American Forage and Grassland Council, but seldom does anyone receive commendations there without growing silage or annual crops. And the forage section is dominated by people who sell hay or build better machinery to harvest hay. Goodness, selling hay is like selling soil. I don't think anyone should sell hay because that translocates carbon from one piece of property to another. Nature's template always recycles the carbon on location. The bison eat and poop—pretty close together. And that means if you're buying hay, you're aiding and abetting a system that depletes the carbon cycle.

One of the biggest carbon translocaters in agriculture is the horse. When horse owners buy hay, they are translocating nutrients. How about that, confinement dairy people, who too often depend on biomass translocation? How about enjoying the sheer ecstasy of building carbon wealth by localizing biomass cycling ? For all the alleged scientific discussions about climate change and carbon trading, who is championing grass? Instead, these experts demonize, marginalize, and criminalize the herbivore for methane excretions. The methane changes considerably when the herbivore eats forage and uses its rumen like it was meant to be: like a four-legged portable sauerkraut vat. And if the grass decomposes, it gives off the same methane as when it goes through the rumen of a cow. So who's kidding who here?

The biomimicry, then, requires moving the animals daily to new ground and away from yesterday's grazing area. By using electric fence or herding techniques (where terrain is rough or labor is cheap) we move the mob from paddock to paddock in order to tightly manage the vegetation carbon accumulation. In short, this is considered lunacy.

All but a handful of farmers and ranchers graze their animals continuously, without moving them at all. They just leave the animals on a field every day all year long. And no one with sense would think of putting chickens or turkeys or rabbits or pigs out on grass. That would be a sure sign of dementia. Confession: here at Polyface we do all the above. Lunacy gone wild.

But the sheer ecstasy of following nature's model is wonderful. First, the logistics: moving the cows every day. The very notion of moving the cows every day is sheer lunacy to most farmers. To them, the phrase "move cows" means an all day affair with three pickup trucks, two all terrain vehicles, two cans of Skoal snuff, a lot of dipping, spitting, and cussing, and maybe you can find them all. To me, that phrase means going out, calling the cows, and watching them follow me down the path or into the next paddock. A few minutes. No problem.

If every day at 4 o'clock somebody called you to a bowl of ice cream, you'd probably follow her too. Except for cows, they like a salad bar more than ice cream. A new paddock of fescue, timothy, white clover, red clover, plantain, orchard grass, dandelion, chicory, and wild carrots is like a bowl of Breyer's butter pecan. I love moving the cows because I know how happy they are to see the new buffet. Farmers whose cows continuously graze never get to see the cows enjoying a new salad bar. Ever. It's the same old same old every day of existence. How humdrum is that?

By controlling the access to different pasture areas, of course, the carbon bank builds up ahead of the herd so that when they enter, often the forage is 2 feet or more high. Sometimes you can scarcely walk through it. We'll put a herd of 500 on 2 acres for a day. When they enter, you can't walk through the forage. Just 24 hours later, a mouse would have to carry a knapsack with lunch just to get across the paddock. Anything that's not ingested is stomped. The ground is literally mulched with shredded biomass that feeds the soil community.

Denying cows access until the grass grows to phenotypical expression more closely approximates natural grazing sequences. The more mature grass, in addition to going through its complete accelerated growth cycle, also begins to turn a little brown around the stem. The way nature feeds soil is with lignified carbon. Green leaves don't fall until they turn brown. Green grass doesn't fall over until it turns brown (called lodging). The browning indicates lignification, where the cellulose turns from protein to starch. Starch indicates energy. You can burn brown grass but not green. The ability to burn is an indicator of energy, or sugar, and that is what the cow's fermentation tank needs. She's not making wine, but it's a similar process.

The sugar is what feeds the soil microorganisms, which are made of protein. Grasshoppers get their incredible jumping energy by eating starch—energy—not proteins. By waiting until the grass becomes mature enough to lignify, we most closely approximate nature's large herd grazing template. Under the typical continuous grazing scenario, every time the grass gets long enough for the cow to nip off a morsel, she does. The grass stays stunted and so do the roots. Over time, the palatable species weaken to the point of death while the unpalatable species, including weeds and brambles, grow unharvested and gain ascendancy in the sward.

This is the way pasture complexion changes over time. It can either move toward better species or toward weeds and brushy species, depending on how close the farmer adheres to the buffalo-wolf pattern. Here again, as weeds and thistles invade his pastures, the typical farmer curses them, as if their encroachment is something coming from outer space. These alien invaders. He religiously applies herbicides to beat them back.

Farmers even get together in political lobby groups to get the government to give them free herbicides to control these invaders. Several years ago our county started a new program offering cost-share monies to farmers to spray herbicides on weeds. How generous. How helpful. What wonderful problem solving.

It never occurs to them to solicit comments from me about how we change fields from weeds and thistles to clovers and grass with grazing management. The very notion that grazing can be helpful and not hurtful is lunacy. After all, the cows are extracting grass. How could this be beneficial, they reason?

The truth is that if the grass gets tall before grazing, instead of the grazing being a negative it's a positive because it restarts the juvenile growth curve. Because the plant is at energy equilibrium, with a nice carbon bank account in its root mass and charged battery in its crown, it can send forth new shoots aggressively and restart the photosynthesis solar collectors. The pruned-off root mass feeds the azotobacter, actinomycetes, and earthworms which in turn make the plant grow healthier than it would without being grazed. If it's not grazed, it just gets old, falls over, oxidizes, quits metabolizing solar energy into carbon, and eventually implodes on itself, kind of like an ingrown toenail. Or a narrow-minded politician who uses force to extract money from the populace to spray weeds because the weed experts say that's the only way to get rid of them.

This daily move of the animals from one spot to another affords the luxury of being able to see them all every day. Most cattle farmers spend half their day driving up and down every hill and holler trying to see all their cows. Every day, our whole herd parades by us on the way to a new salad bar. We can see who's doing well or poorly, or who's been naughty or nice. Why would anyone want checking their livestock to be this easy when instead they could spend half their day looking behind bushes trying to find critters? That certainly sounds like more fun to me. Having them parade by you every day, on their own volition, without being rounded up, prodded, and cowboyed—no, there's no fun in that. How boring.

In pure economic terms, the single biggest advantage of this natural grazing model is that it allows you to reduce hay feeding. For the uninitiated, hay is simply solar-dried forage. Straw is the dried stalk and leaves of a grain plant after the grain has been harvested. Grass does not grow consistently throughout the year; it grows faster sometimes and slow other times. In most areas, it turns dormant either during drought or winter. For those times, farmers usually rely on stored feed in the form of hay to get the animals through the no-growth period.

Of course, making hay is not an easy process. You have to mow it, rake it into windrows so a baler can pick it up, then bale it, then move it to a storage area hopefully under roof. But you're not done, because then you have to take it out of storage and deliver it to the herd where they can eat it. Most farmers brag about how much hay they make. Here at Polyface, we like to brag about how much hay we don't have to make. Even conventional agricultural economists agree that the single biggest component of profitable cattle production is the amount of hay required. More hay, less profit. Less hay, more profit. Very simple math.

Interestingly, in nature, nobody made hay to feed the bison. Nobody today is carrying hay out to the cape buffalo during the dry season. Yes, the animals do migrate, but what if we create mini-migrations by moving them around on our farms? At Polyface, we allow the forage to accumulate in the field and then give grazing access to one day's worth of grass at a time. Tomorrow the herd gets the next spot. And so on through the stockpile until growth resumes and the grass gets tall enough elsewhere to graze. In other words, rather than making hay, we purposely hold back certain areas to let the grass accumulate.

By knowing how much square yardage the herd needs every day, we can easily calculate the number of days' worth of grazing. Just like counting bales of hay in the barn. We just march the herd across the field, systematically grazing the unbaled hay. All we have to do is move a strand of electric fence every day. No diesel fuel, no machinery, almost no labor.

Daniel was down at the sale barn one day and an elderly farmer (most of them are now) came up to him. He had watched us moving the mob across the face of a field near the highway on one of the farms we had leased. It was January. He'd been feeding hay since October. He asked, kind of sheepishly: "Am I right in thinking you all aren't feeding any hay to that bunch of cattle?"

Daniel: "Yes sir (I've taught him to respect his elders). We just move them to another paddock every day."

The farmer scratched his chin a bit and meditated on that. Half a dozen farmers a day drove by in their one ton duallys with hydra-bed, carrying two round bales. Up and down the road. Up and down the road. A constant procession of hay and diesel fuel and labor. Hay and diesel fuel and labor. Finally, he ventured: "Do you all do that o...or...o..."

Daniel: "Organic farming?" The dear fellow didn't know how to say the word, stammering over the foreign language.

Farmer: "Yeah, yeah, that's it. "

Daniel: "Well, we like to call it grass farming. Or beyond organic. This really doesn't have anything to do with organic, but we don't use fertilizers. We let the solar accumulated carbon from the tall grass fertilize the soil."

Farmer: "When are you going to start feeding hay?"

Daniel: "We figure we have about 3 more weeks there. That will put us to the middle of February. Then we might have to feed hay for 3 or 4 weeks."

Farmer: "How do you do that?"

The gratifying part about this whole encounter?—it was one of the first times a conventional farmer ever asked us, sincerely, about what we were doing. Most of them just drive by shaking their heads knowing we're lunatics. So this conversation was a significant step up.

A city fellow came for a farm visit one time during a drought and asked us why our farm was the only green one around. All the other pastures were burned up brown. I quipped: "It's because we get more rain than they do."

He laughed uproariously. "You can't be serious," he hee hawed.

"Oh yes," I said. "If you don't believe me, ask someone."

The reason he was in the area was because a farmer about a mile away as the crow flies was having a potluck shindig and this friend had been invited. He had simply come early enough to slip

25

out and see our farm before going over to the shindig. We finished the impromptu farm tour a few minutes later and he went over to attend the gathering.

Two weeks later he was back in the area to visit the same friends, but came by the farm to buy some meat and eggs. He reminded me about the conversation we'd had two weeks prior and related this story: "When I got over to the potluck, I pulled the farmer aside and told him I'd been over to Salatins and they had lots of green grass over there. Without blinking, he instantly responded 'Oh, they get more rain over there.'"

The sheer ecstasy of being a lunatic farmer includes building organic matter and insulating ourselves against drought. Let me hasten to say that we don't like drought any more than anyone else. But it sure is a lot more fun knowing we're not as susceptible to it and knowing each year we become more and more resilient. These other guys whine and complain and cry and lobby for low interest drought assistance loans.

Nobody in the bureaucracy, and nobody in the legislature, and nobody in the executive mansion even thinks to ask them: "Are you following nature's grass management template to reduce your vulnerability to drought?" That would be a grossly judgmental, unsympathetic question. You see, a farmer has to say we get more rain at Polyface because anything else makes him responsible. As long as he thinks we get more rain, he's a victim and his burned up pasture isn't his fault. But as soon as something can be done to alleviate the situation, he now has to think differently. And that's impossible.

Because our grass stays at energy equilibrium instead of being grazed into the ground all summer, it greens up earlier in the spring and of course stays much greener well into the fall. Every year we begin grazing a month to six weeks earlier than everyone else. In a continuous grazed system, the cows just stay on a field year round. So the grass goes into the winter weak. It survives through the winter and those tender new shoots come up. Wham. Along comes Bossy and whacks them off. Now the plant is weaker. I call this plant infanticide.

The result is retarded growth all season long. Less solar energy metabolized into decomposable vegetable material. Less carbon sequestered. In general, here at Polyface we feed hay only one-third as many days a year as other similarly-sized operations. That's huge. It's worth more than $100 per animal per year.

And not only that, but this biomimicry simply grows more grass, in aggregate, than the norm. In Augusta County, Virginia, the cow-day average (a cow-day is what one cow will eat in one day) is 80 per acre per year. To go through a whole year, the average farm needs 3 acres of pasture (3 X 80 is 240 days) and 1 acre of hay (125 cow-days on one acre is the other 125 days). That's 4 acres per cow for the 365 day year. On our farm, we average 400 cow-days. That's less than 1 acre per cow. Folks, I'm not making this up.

In raw productive volume, then, this historic grazing template spins circles around the conventional model. It's more productive, more profitable, more fun, and more environmentally sound. Every year we watch farmers mowing and making hay and toting it around while we just move a strand of electric fence. It's almost like they are masochists who went to school to find out techniques for making their life difficult and miserable. And these are the folks demanding special concessions and subsidies from the government. Give me a break.

We call the natural system mob stocking herbivorous solar conversion lignified carbon sequestration fertilization. Let's go through this for all the lunatic wannabes out there. First, mob. Yes, we want to mimic the mob, not widely spaced lethargic picnickers out in the field. We want to amalgamate and aggregate enough numbers so that the herd looks like the wild mob, densely packed, eating aggressively and marauding across the landscape in periodic disturbance.

Stocking means it's a managed group on a managed parcel of land. The whole thing is a carefully orchestrated, choreographed arithmetically-contrived plan.

Herbivorous of course refers to herbivores. These critters have the unique capacity to turn forage that doesn't need to be planted, that grows on hills, that doesn't need plowing or harvesting or storage, that self-fertilizes, into nutrient dense meat and milk. It's marvelous.

Solar means it all runs on the sun. Rather than the farm running on petroleum, it runs on real time solar energy. The cows eat, move themselves, fertilize for themselves and receive all the energy they need directly from photosynthesis in the plants.

Conversion describes the changes taking place from sun to plant to animal to nutrient density. Chlorophyll is still by far and away the most efficacious solar collector in the world. And will be for a long time.

Lignified speaks to the browning of the organic material. We want that complex cellulose, not the simple, watery mush of tender proteins. Give me that starch.

Carbon is what this is all about. Converting solar energy into carbon.

Sequestration means we're putting that carbon into the soil. Instead of pulling it out of the soil and putting it into the atmosphere, we're going back to the model that has served the earth quite well since creation to maintain health and equilibrium. We're putting that carbon down into the soil, imbedding it in the vegetable matter and microorganisms that breed and feed on biomass.

Fertilization describes the end and the beginning of the cycle. All of this actually feeds the soil, which then becomes more adept at growing plants that more vibrantly collect solar energy to put yet more carbon into the soil. And this entire cycle depends on the herbivore, both its eating and pooping, to synergize it.

There you have it: mob stocking herbivorous solar conversion lignified carbon sequestration fertilization. If every farmer in America practiced this prehistoric system, in fewer than ten years we would sequester all the carbon that's been emitted since the beginning of the industrial age. It's really that simple. One of the most environmentally-enhancing things you can do is to eat grass finished beef. That sequesters more carbon than soybeans or corn or any other annual. And yet how many radical environmentalists have turned to soy milk and veganism in order to be earth friendly.

But you won't find credentialed experts lining up to encourage the natural system. You won't find farmers lining up to do it. Most farmers think it's too much work. After all, moving cows in their paradigm is a big hairy deal. But when you add up the extra production, extra profit, extra health, extra carbon, extra moisture retention, extra leisure, and extra happiness...well, that's the sheer ecstasy of being a lunatic farmer.

TAKEAWAY POINTS

1. Long grass sequesters more carbon than short grass.

2. Grazing management is the art of domestically mimicing the predator-prey, herbivore-perennial relationship.

3. Grass is more efficacious at sequestering carbon than forests.

Chapter 3

Small is Okay

Before industrialism, farms were localized and seasonal. The ebb and flow of production and activity followed a pattern dictated by local economies, weather, and availability of nearby materials.

Even one of the earliest industries in the Jamestown colony, glassblowing, depended on silica that was discovered nearby. Famous Virginia hams existed because before refrigeration, fall weather in Virginia was conducive to curing. In order to cure a ham with brown sugar, salt, and pepper, the nights need to be cold enough to keep the meat from spoiling until the cure takes. The days must be warm enough to let the juices ooze and suck up the cure.

In the north, fall nights can freeze the hams and thereby shut down the curing process. Farther south, nights don't get cool enough to protect the uncured meat from spoilage. It's a ticklish process, and only Virginia had consistently perfect fall weather to insure good curing. Virginia ham became famous not because Virginians liked ham. Not because pigs like Virginia. Although Virginians do like ham and pigs do like Virginia, Virginia hams became famous due to local weather conditions conducive to curing.

Before cheap petroleum energy, farms depended on their region for inputs and markets, for the most part. Interestingly, Virginia tobacco was the exception to the rule and any cursory study of that plant shows its shortfalls. Of all the plants farmers can grow, it is the most soil depleting. Because it depended on foreign markets, it attracted and gradually encouraged a plantation class of people.

The planter class distinguished itself early on from the more imbedded farming class. The planter class would have endorsed modern industrial farming. They had a voracious appetite for land, slave labor, and exports. Sounds kind of like Tyson, doesn't it? The main difference is that at least when they owned slaves, they generally wanted them in good health.

The planter class represented gentility. Even a cursory reading about their mindset and economics reveals an expectation of recreated English gentility. Place settings imported from England. Wines imported from Europe. Clothes imported from England. The planter class early on knew that tobacco ruined soil. Reading their diaries reveals an incessant emotional struggle with the weed. It destroyed their land but supported their lifestyle of gentility. Caught up in that vicious circle, plantations survived largely by moving west or acquiring other lands nearby.

But even the planting class had its on-farm diversification. Large gardens to grow their own food. A blacksmith. A woodworker to build and to fix buggies and equipment that in that day relied largely on wood. They made their own soap, their own candles, their own clothes from on-site spinning and weaving. Power came from water, animals, or the woods. All of these were locally sourced.

Compare that to today's confinement turkey industry, which started just 30 miles north of our farm in Harrisonburg, Virginia. The only reason the industry started there was because an entrepreneur named Charles Wampler began raising turkeys in confinement. Eventually the breeding program at the USDA research farm in Beltsville, Maryland, developed the double-breasted turkey. By that time, the pharmaceutical industry was up and running to supply cheap medications so that the birds could be kept alive in extremely unhealthy and unnatural conditions.

The entire industrial food system was only possible because of antibiotics for animals and pesticides for plants. Without those two things, these anti-nature production models would not exist and humans would still be dependent on multi-speciation, intricate relationships, and indigenous conditions.

Harrisonburg was not an especially turkey-attractive place. By the 1960s, when the fledgling industry was just getting a toehold, Harrisonburg was not a grain production region. It certainly did not produce antibiotics. It was not even a lumber or steel roofing area, two components needed to build these massive housing facilities. Cheap labor was not available because the local economy was healthy and diversified. The point here is that nothing except one man's entrepreneurial savvy accounted for the industrial turkey paradigm to launch in Harrisonburg.

Today, this industry completely dominates the local economy and community to the point that most people believe it is the local economy. But it has a tainted underside that is worth examining. First, it requires hundreds and hundreds of farmers to grow these turkeys. In the wisdom of the business model, as a vertical integrator, the turkey company owns the hatchery, the birds, the feed, the processing, and the marketing. The farmer signs a contract that requires him to supply a house and labor.

In many cases, since the farmers don't have the money to build a $300,000 football-field-sized house, they mortgage the farm to borrow the house construction money. Often, this is borrowed from the turkey company, thereby giving two income streams to the turkey company: interest on mortgage payments, and turkey sales. This arrangement converts the farmers from autonomous decision-makers to a completely dependent class of people dependent on exports, off-farm inputs, and outsourced decisions.

Suddenly, the farmer's activities are completely ruled by off-farm decision makers. The ultimate outsourcing. It's one thing to outsource raw materials, markets, and accounting. But when virtually all decisions are made in board rooms located outside the community, very few locally-appropriate decisions will be made. I've guest-lectured in communities devastated by this model.

Farmers signed over their farms to join this scheme, only to lose their farms 20 years later when the turkey company decided to change course.

To add insult to injury, these farmers now become major blights in their community. Neighbors suddenly must either suffer silently in the stench created by these houses, or suffer belligerently either in courts or from vigilante phone calls—or bad mouthing in the community. The backlash became bad enough for the American Farm Bureau Federation, lover of everything industrial, to push through Right-to-Farm laws in most states of the U.S. I call these Right-to-Stink-Up-The-Neighborhood laws. The assumption that fresh country air must stink is a direct result of poor farming.

Every day some 100 train car loads of grain come into the big company-operated feed mills that supply the poultry houses in the area around Harrisonburg. In the houses, this feed is converted to meat and poop. If the industry were scaled to be dependent on ecologically-sensitive indigenous production capacities, it would not exist in Harrisonburg. It would be centered in the Midwest, where the grain is grown. Today much of this grain is coming from Argentina through the Atlantic seaboard.

All of this poop has to go somewhere. Essentially the poultry industry has turned Harrisonburg and its environs into a giant toilet. The direct water pollution connection has prompted the industry to institute Best Management Practices to create protocol for dealing with all the poop. Virginia Tech's Extension Service, in collaboration with the USDA, began promoting feeding the poop to cows. The owner of the Harrisonburg slaughter house that we use and now co-own used to buy Shenandoah Valley beef to sell in his retail store. But once this new scientific feeding method became widely adopted, he quit because "I got tired of walking in the chill room and the meat smelling like chicken manure."

Of course, all the bureaucrats conducting scientific research into this method cranked out press release after press release extolling its virtues and proclaiming that it had no effect on the meat. Oh dopey me, why would anyone think that diet has anything to do with meat quality? The notion is absurd. So today you can drive right down the road in our community and see feedlot beef eating rations of chicken poop. They mix it with some silage and a little molasses and the cows eat it up: poop, decomposed chicken carcasses, et. al. And I'm the lunatic. Right?

This poop is a real problem because it has to spread out over large areas in order for the soil to metabolize it. It's not balanced, so it oversupplies certain nutrients to a toxic level. The search for new ground continues. Fortunately, the spike in petroleum prices came in the nick of time to create new interest in the fertilizer value of the poop. It's being trucked clear out of the Valley to get it onto new ground. A pelletizing industry has now attempted to dehydrate and pelletize the poop. This allows it to be bagged for homeowners to apply to their lawns—lots of new ground there.

Pelletizing also enables it to be handled with augers and mixers so it can be blended into other fertilizers. Now we hear about a new plan: biodiesel. Turn the poop into fuel. Of course this is all being prototyped at taxpayer expense, with under-the-table sweetheart deals to the planters...oops, I mean the corporate moguls who will eventually release Initial Public Offerings in their lucrative ventures developed and researched by the gullible taxpayer who has to live every day in the stench and dust of fecal concentration camp factory poultry houses.

After turning farmers into serfs, the industry needs lots of labor to process these birds into turkey ham, turkey salami, turkey franks and the occasional whole turkey for Thanksgiving. Lots of labor. The plants are dangerous, dirty, unhealthy places to work. Neighbors don't want to work there. So the industry goes outside the community again. Way outside the community. Even outside the U.S. To Mexico, to be exact. Now I don't want this discussion to get into a brouhaha about undocumented workers, illegal aliens, or protectionism.

But I do think it's important to understand, from a community standpoint, what this does. I'm keeping this confined to the Harrisonburg area because that's my area and one I'm very familiar with. Be assured that this scenario has played itself out in communities all over America. This story is the backbone of the industrial food system. This foreign work force floods the community. So much so that just recently Harrisonburg had to build a new school because more than 30 percent of the classroom space was occupied by English as a Second Language (ESL) training.

Although this is not an armed takeover of a community, it is similar in that the rapid and coordinated influx overwhelms indigenous norms. Now this sleepy little heavily-Amish influenced community has Hispanic street gangs. Machete-wielding criminals must be imprisoned. Interpreters must be provided for their court hearings. Nobody loves other cultures more than I do, but nature moves toward balance. I like trees and I like ponds, but you don't see trees growing in the middle of ponds. Unless, after many years, the pond silts in enough to support trees in what is no longer clearly a pond. Natural succession is inherently gentle. Unless it's a volcano. I'm sure all of us would love volcanoes in our communities.

This industry has received a few shocks over the years. The most recent was an outbreak of avian influenza that resulted in the destruction of 1,000 tractor trailer loads of chickens and turkeys, but primarily turkeys, in the immediate Harrisonburg area. A task force of 70 federal veterinarians spent nearly two months in the epizootic containment program. Of course, all of those vets needed brand new F-150 4-wheel drive pickup trucks to tool around from farm to farm, collect samples and protect the industry.

Actually, the jury is out about what really happened to those birds. Because it was a government-ordered eradication program, taxpayers indemnified the birds, to the tune of the better part of a billion dollars. The official story is that they were all either incinerated or landfilled. But if you talk to insiders, they say most of them were processed as usual and went right into market channels. When these kinds of things happen, the industry circles the wagons so tightly, and the money is so big, that journalists either can't or won't penetrate the official press release.

During this time, two of the federal vets wanted to come out to see our farm and I was more than happy to spend some time with them. One came one week and one another—the vet task force rotated every so often to prevent away-from-home-fatigue and perhaps to keep anyone from knowing too much. Newcomers can usually be manipulated by the industry vets. Anyway, each of these men told me that every vet on the task force knew that the reason for the outbreak was too many birds in too close confinement in too tight a geographic area...BUT, if any of them breathed that to the media, he would be fired within 24 hours. There's your government report and science for you.

In fact, each of them said that here at Polyface, we were considered Typhoid Marys because our pastured chickens commingled with red-winged blackbirds, which could take our diseases to the science-based environmentally-controlled poultry houses and threaten not only the industry, but the entire planet's food supply. After all, if the poultry industry goes down, the world starves. We all know that. I mean, what would we do tomorrow if we couldn't get chicken McNuggets? What an awful death that would be. Death by McNugget deprivation. Horror of horrors. Jodie Foster, where are you when we need you?

When I asked these vets what would happen if our farm did not submit to testing or cooperate with the feds (we know farmers whose tests have been tainted at the laboratory—I don't trust these government officials as far as I can throw a bull by the tail, and that's not very far), their answer was sobering indeed. Both vets looked at me, dispassionately, and explained that on the day these credentialed experts released a report saying Polyface would not cooperate, the community would rise up and crucify this Typhoid Mary. "You can't win that. The community trusts the government report," they said. Whew!

What's the bottom line here? We've talked about train car loads of imported grain, mountains of poop, massive foreign worker issues. The bottom line is that in my region, to disparage the poultry industry is akin to assaulting America. Good patriots agree: not only is this poultry industry good for our local economy, it is in fact the foundation of our local economy. And to suggest anything else is to hate your neighborhood. If you suggest we may have been better off without it, you're in favor of massive unemployment, bread lines, and homelessness. In fact, you're a lunatic who must be silenced. The Democrats and Republicans are equally dependent on the industry because it represents jobs. We can't do anything to jeopardize jobs.

Okay, let's talk about these jobs. Twenty years ago the industry was being chewed up with carpal tunnel syndrome, or repetitive motion disorder, workmen's compensation claims. The direct causal link between these processing plant jobs and repetitive motion disorder was well established. You just can't stand all day every day on a processing line, making a 2 inch cut on a turkey carcass, without damaging your tendons from that repetitive motion.

What to do? Easy. The industry lobbied the Virginia General Assembly and by law eliminated repetitive motion disorder from the list of workplace illnesses. Simply by fiat the industry exempted itself from these claims. Nice industry-government collusion. That takes care of the problem. Didn't change working conditions. Didn't change the jobs. Didn't change the illness. Just say it ain't so and keep cranking out McNuggets. Of course, lots of these folks don't have health coverage. So to deal with that, the country must pass another corporate welfare sweetheart deal: taxpayer-sponsored health care. The Democrats being stupid are just as culpable as the Republicans bailing out industry responsibilities. Oh, it's a tangled web.

Nothing about the poultry industry generally and turkey industry particularly, as illustrated by Harrisonburg, is local. Most of the turkeys are not sold in Harrisonburg. Their feed does not come from Harrisonburg. The labor to process them does not come from Harrisonburg. The poop can't be handled in Harrisonburg. The whole deal, top to bottom, has nothing to do with indigenous resources, markets, or labor.

And yet, for all this, farmers are still lining up to borrow money to build poultry houses, viewing industrial poultry as a panacea and an opportunity to hold onto their farms. It pollutes the community, upsets the neighbors, clogs the schools and prisons, and turns farmers into serfs. Amazing. And the industry just keeps on building and growing as if in the perfect world, every square foot of the Valley would be covered with a confinement poultry house and we would become a septic tank instead of just a toilet.

After all, bigger is better, right? Growth is always good, right? Remember, cancer is growth. Growth without responsibility is not healthy. Just so we can all be on the same page, let me list a few things we'd like NOT to grow:

Disease
Pollution

Illiteracy
Jails
Murders
Bank robberies
Assault and battery
Poverty
Sickness
Divorce
Pornography
Drunk driving
Hunger
Taxes
Attorneys
Bureaucrats
Military industrial complex
Processed food
Starbucks
McDonalds
People magazine
Petroleum
Flies
Gullies
Pesticides
Genetically Modified Organisms
Irradiation
Cloning
High fructose corn syrup
Deserts
Liability insurance
Nursing homes
Government jobs
Wall Street
Unscrupulous bankers
Cattle feedlots
Unloved children
Two-income dependency
Dependence on the government
New York City
Centralization of food

Corporate farms
Plowing
Insane asylums
Drug use
Unemployment
Prejudice
Welfare
Obesity
Type II diabetes
Atmospheric carbon
Socialists
Monsanto
Globalists

Okay, that's probably enough for now. The point is, in normal conversations the assumption is that growth is good. I disagree. Only good growth is good. We could all have a healthy discussion about what good growth is. Even to question a plan for bringing jobs to the area or building another highway or subsidizing another corporate office complex puts you in the lunatic category. The problem is that we have short memories and see with myopic vision.

Seldom is the assumed normal way the only way. The only way to deal with congested highways is to build more highways. Right? NO! If you incentivised businesses to stagger their office hours, it would spread out road use and eliminate rush hour. Or perhaps businesses could be incentivised to encourage telecommuting. Or perhaps if small businesses were not prejudiced with government policy, these goods and services would be more spread out across the landscape. We could go on in this vein for some time, but almost every problem has a context and several paths toward solution.

Compare all we've talked about with the Polyface pastured poultry model. First, it's seasonal. We aren't burning propane to keep chicken houses warm. When it's hard enough keeping all the people warm in a community, isn't it strange to be keeping chickens warm? We let the season dictate the production time frame. Over the years, many patrons have begged us to raise meat chickens (broilers) through the winter. We have steadfastly refused. First, it would take lots of energy to do so. Secondly, we want a break.

This is not a factory, where raw ingredients come in the front door and processed packages come out the back door with no relation to the ecological umbilical that ties us all to the earth. On the contrary, our farm is imbedded in the local mores. That includes seasons, resources, labor, energy, markets. If we exceed or discount any of these constraints, we have foisted upon the community new extraneous problems. We can sleep at night knowing we haven't stunk up the neighborhood or nitrated the groundwater to kill a neighbor's child with blue baby syndrome.

Our grain comes from local farmers who do not use genetically modified organisms. And if we could, we'd even get them to grow open pollinated varieties so they could save their own seed and eliminate dependency on far-flung seed companies. And these seeds would become acclimated, each year, retaining some genetic memory of the area to create nativized genetics. A new crop of indigenous DNA. Wouldn't that be cool? Beyond that, our chickens are eating a fair amount of forage. The turkeys eat up to 40 percent of their diet in grass, which tremendously reduces the grain component per pound of meat. If everyone put their turkeys on pasture, just think how much grain, tillage, trucking, and petroleum we could save.

But what about off season? That's what freezers are for. And they are a lot cheaper to run in the winter than in the summer. In areas where the winter would naturally shut down pastured poultry production, the seasonal cold makes storing in freezers quite cheap. As ambient temperatures drop, the energy requirement to maintain freezing temperature is less. A lot less than trying to keep birds warm in the winter. And although the body heat generates significant warmth, the birds must eat extra feed to have enough calories to give off heat.

We process right on the farm. No late night interstate travels, spreading feathers all over the countryside. What a strange thing, to process the birds right where they grew up. Shouldn't they all have to go to a megalithic concrete monument to the stupidity of man in order to be readied for eaters?

Our processing crew is not single-tasked, but multi-tasked. Nobody should kill animals every day. It's not good for your psyche. I've certainly done my share of butchering, but I wouldn't want to do it every day. Just like a lot of things. Variety is not only good for the emotions, it's also good for your muscles and body agility. The famous quotation attributed to Henry Ford sums up the factory mentality pretty well: "The worst part about this is that I have to employ a whole man when all I need are his hands." Very funny, Mr. Ford.

A couple mornings a week our crew may process chickens. In the afternoon, they might make hay or move pigs or pull weeds in the garden. We don't process every day, we don't process all day, and processing poultry is not all we do. This completely changes the human-ness of the process.

But doesn't our system take way more land than the efficiencies of confinement factory houses? Not at all. In our system, the birds are out on grass, dropping their poop and eating grass plus grain. In the confinement houses, their grain has to be produced somewhere and their poop has to go somewhere. Even if our birds didn't eat any grass and consumed the same amount of grain, the land required to grow the grain would be the same per pound as it is in a confinement house. No land difference there. Everyone needs to understand that radiating out from every single confinement animal operation, whether it be poultry, pork, beef, dairy, or guinea pig, an entire unseen land base supports it. You don't see the corn fields. You don't see the corporate offices. You don't see the manure hauling trucks and the acres on which the manure is spread. Our pasture based model actually takes less land than the industrial model.

But how can you feed the world? I think we just answered that. The land requirement is actually less. More acute is my presumption that globalist agriculture should simply not be practiced. We would actually have a stronger local economy, a stronger local social structure, a stronger local ecology, if Harrisonburg did not depend on exports to maintain its poultry empire.

Several years ago I had the distinct privilege of attending and speaking at Terra Madre, the international Slow Food convivium in Turin, Italy. A virtual United Nations of sustainable agriculture and food-craft artisans, the 5,000 delegates came from about 130 countries. I enjoyed extremely colorful tribal costumes and great presentations about indigenous food production from all over the world.

I made a point to attend every presentation by an African delegation that I could squeeze in between my other responsibilities. In every single one, the speaker said their community could feed itself quite well, thank you. But cheap western (primarily American) subsidized imports, dumped in those communities, dislocated the indigenous food economy. It created a dependency mentality and displaced local producers and vendors.

A middle-aged fellow came by the farm recently who had spent several years in South Africa working with USAID. He had recently resigned and I asked him why. He said he grew weary of the displacement the foreign aid was creating in the countryside and began pursuing answers. Finally he reached a high official who expressed incredulity to him: "You don't get it, do you? Don't you realize USAID exists to create dependency on U.S. products in these countries?" That's a strong statement.

All you have to do is travel and talk to anyone except the Wall Street press, and these stories come to light. If you look at what Allen Savory and Holistic Management International have done in Zimbabwe, you realize very quickly that U.S. foreign aid is completely misguided, unnecessary, and destructive.

After Chernobyl blew in Russia, the radioactive cloud drifted over Belarussia, which was the dairy region of the old Soviet bloc. The radioactivity settled in the mammary glands of the cows, but not in the meat. The people were starving. The U.S. sent millions of dollars of humanitarian aid over there. A year later, their Secretary of Agriculture along with their equivalent of our Speaker of the House and a couple of other high ranking officials came to our farm for a visit. They arrived in two sleek black limousines and were clearly cultured upper crust, wonderful and delightful folks. After touring around our farm, here is what they said (through the translator): "The day the foreign aid was deposited in our bank, every hotel filled up with U.S. corporate salesmen from machinery companies to seed to chemical companies. All that money was spent on things we did not need, things we could not fix, things we could not afford to put fuel in. If we had known about your kind of farming, we could have put in

42

water systems, fence systems, and gone to a pasture-based system and fed our people and had enough left over for export."

Talk about the sheer ecstasy of being a lunatic farmer! A friend does missions work in the non-tourist part of Kenya. Recently they built an orphanage and day school there. A local tribal chief came to the U.S. to speak to some of the churches who supported the mission work. Since he was in the area, my friend took him to see an industrial chicken house and then came over to our farm. Nothing at the industrial chicken house was applicable to his people, he said. But when he came to our farm, he became animated with excitement.

"Yes, we can do this. Yes, this will feed our people. Yes, this will stop chicken diseases." He realized how appropriate and doable this simple infrastructure and moveable system could be. I gave him my *PASTURED POULTRY PROFITS*. When you have a model that works across socio-economic lines and across cultures, you know you have something. I call it truth. Most folks around here call it lunacy.

The point is that clever, hardworking and skilled folks live in every community of the world. Every community is endowed with air, soil, and water. I don't know what it is in the human spirit that makes us want to meddle. Apparently it's always easier to tell somebody else what they should do than to do what is right ourselves. Be assured that when the U.S. exports advice, it's no better than the industrial fecal food it exports. The U.S. has almost twice as many people incarcerated in prisons as we have people farming. At least prisoners receive their own square on the Census Bureau tabulation; farmers don't even merit that. But despite this lopsided statistic, the USDA thinks we're agriculturally superior and other countries should adopt our methods. Is this ratio really a sign of strength? Or a sign of collapse? Or a sign that we don't have a clue as to what we're doing? I don't think it's a sign that we should be teaching.

As an author writing for farmers, I would have twice as big a patron base if I wrote a book *HOW TO SURVIVE AND THRIVE BEHIND BARS* as I did when I wrote a book titled *YOU CAN FARM.* I think the USDA budget should be tied to the number of farmers. Of course, if it were up to me, we wouldn't even have a USDA. No government agency has been more successful at annihilating its constituency. While certainly some wonderful people work for the USDA, the outfit is rotten to the core, in my opinion. And good people always find places of service. They don't have to work for rotten outfits.

Certainly our localized, multi-speciated, pasture-based system requires more farms, more farmers, and more people scattered out across the landscape. But what is wrong with that? I can think of a lot worse situations to find myself in than being cooped up on a farm (no pun intended). I may not make lots of money, but I sure have a great office. Plenty of people cooped up in Dilbert cubicles working as a cog in a multinational corporate machine would give their eye teeth to be stuck on a farm if they felt like they could make a living on it. And that is partly what this book is all about. You can make a living on it, but you'll need to think and act like a lunatic when compared to the presiding paradigms.

I think repopulating the countryside with loving stewards is a great aspiration. I think it might even be a good national security policy. So would populating our homes with lovers of domestic culinary arts. More on that later.

What a joy to know that our farm isn't dependent on foreign currencies and foreign resource streams. That it works right here, or anywhere. That it can empower a Kenyan tribe to feed themselves rather than make them dependent on my anti-community empire. That, folks, is the sheer ecstasy of being a lunatic farmer.

TAKEAWAY POINTS

1. Growth can be cancerous.

2. Most communities, even in developing countries, can feed themselves.

3. Localization is about being connected to the ecological umbilical.

4. U.S. foreign aid usually does more harm than good.

Chapter 4

Crooked Fences

I love crooked fences. On our farm, we don't have straight fences—except on boundary lines. Otherwise, all the fences are crooked. Now if that's not advocating lunacy, I don't know what is.

Why would we want crooked fences? Because the land lies crooked. And the land should define how the fences run. Rolling and hilly land certainly has more topographical definition than extremely flat land, but ultimately all land lies unevenly. Topography is the most defining feature of a landscape. It defines the way water runs, vegetation, and access.

The three great environments are open land, forest land, and riparian. Certainly some areas are hybrids, like savannah or swamps. But these combinations notwithstanding, in farming we're generally dealing with fields, forests, and water. Defining those with fences is like making raised beds in a garden.

Anyone who enjoys gardening knows the frustration of toddlers or little children running through the plants. Sometimes it's hard to tell what is garden and what is lawn. The clearer the borders between walking areas and non-walking areas, the easier it is to incorporate children. Building up garden beds with boards

46

creates nice physical barriers that protect against marauding children. The easier children can see the borders, the more comfortable adults can be with their help in the garden. Or with children playing near the garden.

Fences act the same way on the farm's landscape because each of these environments requires a different management system, just like the garden beds versus the lawn. Think about the prettiest gardens you've ever seen. They probably curved around the contour of the ground. Circles and rounded edges. Yes, some are in square blocks, but generally tear drop shapes and curves make a more interesting and appealing shape.

That is what we're doing with the farmscape. Rather than disregarding or even fighting against the land's flow, on our farm we enjoy letting the topography define these physical lines of demarcation. We let the land speak to us, as it were. The land is the dominant voice here, not us. And that attitude definitely puts us in the lunatic category.

After all, the overriding opinion in mainline farming is that the farmer is the overriding voice, not the land. As a result, farmers in our area—just like every area—build straight fences and square corners. It's all part of forcing the land to conform to our desire, rather than conforming our desire to the land. Maybe it goes back to playing with blocks as a child. Squares are imprinted on our brains and we look at everything through the prejudice of 90 degree angles.

What's the problem with this? The most critical issue in rolling country such as we have here in the Shenandoah Valley is that arbitrarily square or rectangular fields necessarily incorporate marshy or extremely steep areas that don't partner well with other areas. For example, we might have a field that is primarily homogeneous, but includes two areas, probably toward the corners, that are either too steep or too wet to drive on much of the time. But since these areas are within the confines of the fence, we risk life and limb driving over the steep area and every couple of years create a new set of deep ruts bogging through the low areas.

And since it's all in the same field, we force it to be handled the same way. The steep area, the wet area, and the intermediate area all get the same treatment at the same time. When our family came to this farm nearly half a century ago,

Dad's rule of thumb was that if you didn't feel comfortable driving on an area because of steepness, it should be abandoned as field. That means it should be forest, or orchard, or vineyard—something that doesn't require hay mowers, manure spreaders, and hay wagons. As soon as you look at a farm from that perspective, you begin making crooked fences.

Most farmers around here don't even fence out their woodlands. Some are beginning to fence out their waterways, but usually that's only done with government cost share money through the Conservation Reserve Enhancement Program (CREP). The problem with CREP is that it builds straight fences that absolutely ape the landscape. We lease a couple of farms that have participated in the program and the square corners along with straight lines couldn't be more inappropriate. After all, a government program's technicians are prejudiced by the same thinking as the private sector.

Animals never walk across the landscape in a straight line. They meander just like creeks. They tend to walk on the contour because it's easier than going up and down hills. In fact, if you laid out your fences according to the traffic pattern of cow herds, you'd be in the ballpark as to where the fences should run. What happens in CREP is that the fences run up to the head of a damp swale, for example, the seepage that begins a stream, to a boxed, square fence enclosure. Or exclosure, in this case.

The fence ought to go around that swale head in a gentle curve. The cows, instead of being able to walk on the contour around a gentle curve, must walk to the corner, make an oblique turn, walk to the other corner, and make an oblique turn. These oblique turns denude the ground and create erosion areas. Then the corner posts rot out and fall over and the whole thing goes to pot.

Imagine how a river meanders through a meadow. What CREP does is come out a certain distance and build a straight fence, which of course sometimes puts huge areas of extra land into the CREP area that could certainly be used for grazing. Oh, I forgot. Grazing is bad for the land. The way most farmers graze, yes it is. But it doesn't have to be. Once CREP builds the fence, you have an arbitrary line that has nothing whatsoever to do with the topography. However, it preserved a nice square field.

Amazingly, nothing about CREP encourages or teaches proper grazing management. Mainline conservationists are just as narrow-minded as mainline industrialists.

You could make the argument that at least CREP keeps the farmer from plowing near the river. My rebuttal is twofold. First, the fence is not far enough away to protect the land from erosion in a flood if it were plowed. To do that, the fences should be placed much farther away than they are. Second, if you practice good grazing management which builds healthy sod which builds healthy soil, that can be done right up to the water's edge without any deleterious effect. Why waste all that land, which in many cases is the most productive on a farm, to satisfy some arbitrary notion of conservation?

The fact that so much land is going into CREP does not indicate a new awareness of soil conservation. It indicates a lack of profitability in farming, so the only stable alternative is to receive a government paycheck for land that's not being used. It's a sad state of affairs when the farmer's steadiest paycheck is the one he receives from a government program, whether it's a corn subsidy or a conservation easement.

To speak against conservation easements is sacrilege in this time of environmental awareness. Like all government programs, the good does not outweigh the bad. Here's an example. We recently leased a farm that had participated in a CREP arrangement that included 120 acres of the 360 acre farm. The CREP engineers came in and designed a system that would fence out a couple of small springs and their downhill swales in exchange for a piped water system.

They bored several wells but all came up empty. Finally they decided to go into the bottom of the strongest of these seepy swale areas and trench in a perforated pipe. They dug out a trench maybe 4 feet deep and 4 feet wide, filled it with rock, but imbedded a perforated pipe in that rock trench. Water flows through the rock easily, then into the perforated pipe. The perforated pipe drains into a buried concrete cistern. Up the hill a couple hundred yards they installed a sophisticated sun-tracking solar panel to power a pump.

The pump sent the water from the cistern to two larger buried cisterns on the top of the hill. The gravity water flowed from these cisterns out to three strategically placed 4-hole frost-free Ritchie drinkers. Remember, this is three drinking spots with only 4 holes per station for 120 acres. The drop in the pipe, from cistern to drinkers, was so slight that at best the water would only run in at a measly 2 gallons per minute. The drinker container held only about 20 gallons. On a hot day, one cow can easily drink 5 gallons in two minutes. Add three more cows to the other three holes, and you have an empty drinker in four minutes.

That meant that anyone who was grazing that 120 acres could not practice rotational grazing because the water flow was not enough at any one point to satisfy the herd that would be there. Continuous grazing was the only option. So here we have a conservation program designed to keep the farmer from practicing the kind of grazing management necessary to build soil. The installed infrastructure forces the farmer to destroy the soil if it's used the way it was designed to be used because the only way to use it would be to continuously graze, which we've already established is a land deteriorating system. Bottom line: the conservation program designed and installed an anti-conservation infrastructure. So much for governmnent help.

To add insult to injury, in the first year, the seep went dry. The farmer had to haul water to the cows to keep them alive. Here was a $100,000 taxpayer-funded conservation project that not only necessitated anti-conservation farming, but wouldn't even keep up with a few cows in the first year. And the fences created erosion at the square corners. To reduce erosion around the drinkers, they poured huge concrete pads out several feet. Of course, that meant all the manure and urine that collected on those either evaporated or ran off to the edge in such quantities as to be toxic to the surrounding soil.

The following year, the landlord leased the entire farm to us. The first thing we did was rebuild a small pond up on a high area and gravity feed down to a cistern where we installed a high pressure one horsepower pump and pressure tank. Then we trenched in a couple miles of pipe with 24-inch diameter access holes every few hundred feet. Each of these access holes has a valve.

This allows us to use a portable water trough. Moving the water trough enables us to fully utilize the manure and urine that naturally collects around a watering point. High impact points, if they are moved around, can become integral components of a fertility program instead of entry points for pollution or translocation points for lost fertility. Placement and timing are everything, whether you're planting carrots or grazing cows.

The bottom line is that we spent only $17,000 on the pond, the pipe, the cistern and the pump. It covers the entire 360 acres, not just the 120. And we've got water up the wazoo. We can look at that pond anytime and know whether we still have 700,000 gallons or 200,000 gallons. No surprises. And everywhere on the whole 360 acres we have wonderful 10 gallon a minute water with great pressure. That's enough to handle a 600-head herd at any spot. And yes, we move the herd every day. And yes, weeds are leaving, the soil is building, earthworms are waking up and copulating. It's all marvelous.

When the landlord saw our system compared to her malfunctioning system that she paid thousands of dollars into as her part of the government cost-share, she felt betrayed by the government conservationists. She called the head design engineer to come for a nonconfrontational walk-about to perhaps incorporate some of our ideas into future projects. He wouldn't come. After all, what can a government expert learn from a peasant farmer?

Polyface now leases several farms in the area. All have come as a result of land owners wanting us to heal the land like they've seen us do here at our home farm or at another farm that we lease. We've installed many water systems. Each one is customized to the terrain and needs of that particular farm. I can scarcely describe the emotional high, the exultation, that I feel when I see water pouring out of a pipe a mile away, out in the remotest corner of the farthest field. It's sheer ecstasy. Water is the beginning of everything, and until you have it, you can't grow anything.

Besides some preconception, or fetish, about squares, the reason most farmers want square fields and square fence corners is because they don't trust electric fence. I never cease to be amazed at the incongruity of farmers quickly adopting some technologies but summarily refusing to adopt others. It probably has to do with instant gratification. Applying petroleum fertilizer boosts yields or turns pastures green almost overnight, creating a dramatic result. When applying pesticide creates millions of dead bugs in an hour, that's power. I was on a forage and grassland bus tour one year and we went by a farm that had herbicided a pasture to plant no-till corn. Everybody on the bus was euphoric over the dramatic burn-down. "Wow, that's awesome. Look at that kill. Isn't it beautiful?" they all genuflected.

The brown, dead grass lay inert on the soil, spring sun welcoming it to partake of fresh solar energy. But it was dead. Nobody was interested in whether or not corn needed to be planted. Or whether corn was appropriate in that spot. I'd have been exhibit A in lunacy had I questioned their euphoria, their chest pounding over what technology could do. In one day, with a few gallons of elixir, it could shut down all the mitochondria, the stomata, the chlorophyll. "Aren't I great?" they seemed to chant.

Many Americans are baffled and indignant when radical Moslems chant "God is great!" as they blow themselves up. I wonder if the euphoria over herbicide's power is any different. In any case, it's a strange god that wants you to blow yourself up. And it's a strange god that glories in good herbicide kills. If a bevy of virgins awaits the radical Moslem, I wonder what awaits the herbicide applicator—intergalactic spray rigs with 100 foot diameter nozzles? Remember, I'm the lunatic for questioning.

Electric fence has been around for a long time. And it's extremely dependable. But farmers don't trust it. They want to build physical fences; multi-wire, heavy, expensive, lots of posts, monstrosities. For boundaries, yes, that's great. But for internal, who needs that? It's just a bunch of extra maintenance. A fence strong enough to physically control a cow or sheep is too heavy for curves and topographical design. Posts eventually lean into the turns as gravity runs its course.

At Polyface, we build one or two strand electric fences for all of our internal fences. Little short posts and aluminum wire so light you don't even need braces at the ends. Just use a little bigger post for corners and ends and keep going. These fences can handle turns and curves easily without bending. And they don't impede wildlife. A farmer asked me for a day of consulting on his farm. He wanted to show me his brand new woven wire fences. Somehow a young deer had gotten into a field, probably through an open gate. When we walked through, of course, the deer panicked. The fence had been built strong enough and tight enough that the deer could not get out. It never did get out, but just kept bashing into the fence as we walked along. The farmer was merely amused. I assume it escaped through an open gate after I left.

The lighter our footprint, the better. Why don't farmers exult in less obvious infrastructure instead of more obvious? The bigger the impact people have on the landscape, the more hubris kicks in. I think a little humility might be better. At Polyface, we just keep putting in our little almost-invisible electric fences and the neighbors wonder what planet we came from. Electric fences allow us to cheaply adhere to the topography as we protect fragile areas and create lines of demarcation between the environments.

Another huge reason for topographical fencing is to create homogeneity in particular fields. A southern slope, for example, tends to dry out faster. The drying out means it often has shallower soils because less moisture means the vegetation browns down quicker in droughts. As a result, over time, since vegetation is what grows the soil, these soils are not as deep as soils on north slopes or in swales. Southern slopes do green up earlier in the spring, so they tend to be grazable or plantable earlier.

If a field, therefore, includes both heavily southern and heavily northern aspects, one grazing or planting scheme over the entire area will be an inappropriate fit. For example, if you're planting corn in the spring, the day the southern slope is ready, the northern one will still be too cold and wet. By the same token, the day the grass is ready to graze on the southern slope, the northern slope is too immature.

On the other hand, in a drought, the northern slope will hold moisture much longer and continue growing. The swale even more. By running fence lines topographically, then, we can lump similar land areas together for more appropriate management. Or said another way, it reduces the risk of mismanagement. The bigger the field, the bigger the chance of mismanaging some of it. This is of course counterintuitive to a paradigm that applauds bigger parcels, bigger machines, and one size fits all.

Crooked fences make more edge. Edge effect is what wildlife biologists call the flora/fauna high diversity zone where field and forest, forest and water, or water and field intersect. Most animals and even many plants need a little bit of two environments in order to live. They may sleep in one environment but eat in another one. Fox squirrels are a good example. The reason you only see little gray squirrels in deep forest is because the much larger fox squirrel needs two environments. He lives in the forest but buries nuts in the field. A straight fence along one of these edges reduces the linear footage of edge. A zig-zaggy edge creates much more linear footage for the two-environment edge.

Running a fence around a pond protects the edges from trampling by livestock. Trampling not only denudes the soil and muddies the water, but it also over time pushes the soil into the pond. Gradually, the pond fills in and doesn't hold water anymore. From a sanitation standpoint, having the cows lounge in the pond not only dumps in harmful manure and urine, but also dirties the water for the cows to drink. If you drank out of your toilet bowl, you'd probably have worms too. But thousands of farmers give their cows unimpeded access to ponds, only to dose the cows up on parasiticides and wormers later on.

Why not fence the cows out in the first place? Because it's a pain to put a square fence around a circular pond. And big physical fences are hard to build. It's a lot easier to let the cows soil the water and fill up on parasites so you can have sick animals and patronize the pharmaceutical industry. Of course, to do all that you have to run them through the head gate and threaten life and limb. But that's what farming is for, after all. According to conventional thinking, farming is supposed to make life difficult and tedious.

When the cows trample in the pond edges, the cattails, sumac, pond grass and other hydrologic vegetation can't grow. If it can't grow, the frogs, newts, and salamanders that normally live in that amphibian zone don't exist. Wild waterfowl can't build a nest because no vegetative protective cover exists. But why would anyone want the ecological stability of all this vegetation and wildlife when they could have an eroding, sterile bank instead?

In the end, electric fence is for lunatics. If Grandpa didn't have it, then I won't either. And since we've been making fields square for two centuries, we're going to keep doing it. That's the thinking of the conventional mind. Meanwhile, I'll enjoy fences that twist and turn; fences that are cheap to build and maintain; fences that don't impede wildlife; fences that are barely visible on the landscape. In short, I'll enjoy the sheer ecstasy of being a lunatic farmer.

TAKEAWAY POINTS

1. Fences should adhere to the topography.

2. Electric fence is cheaper and simpler to build than heavy physical fences.

3. Slope aspect defines vegetation and seasonal growth patterns.

4. Government conservation programs are not holistic.

5. Good grazing management would eliminate the justification for most conservation programs.

Chapter 5

Water Massage

Land management's marriage partner is water. The two go hand in hand because nothing bears more heavily on vegetation and opportunity than water. It's the force that dominates what a landscape looks like. Topography undulates. At any given latitude the average temperature is in the same ballpark. Add water or withdraw water and you have a major change in appearance.

You would think, then, that water management, or what I call massaging the landscape with water, would be right at the top of the farmers' list. If the farmer has a priority list, either written or subconscious, acquiring, storing, and dispensing water should be up there at number one or two.

Believe it or not, it doesn't even register for the vast majority of farmers. Here is the priority list as I see it for American agriculture:

1. Watch the weather channel and then complain.
2. Drive a good looking pickup truck.
3. Drive a properly colored tractor—blue, green, or red.
4. Watch the weather channel and then complain.

5. Do some farmin'—plow something, anything. Just plow.

6. Pass the test and certification to apply pesticides and herbicides.

7. Watch the weather channel and then complain.

8. Have the biggest: corn, cow, lactation. Something. Anything. Bigger.

9. Build a straight fence—somewhere, anywhere.

10. Watch the weather channel and then complain.

11. Go over to the neighbor's, lean against the edge of the pickup truck, and complain about the weather.

Did you happen to notice how often weather and complaining were on the priority list? Weather has to do with water. Complaining sure solves problems. When I say weather, let me be perfectly clear that I'm not talking about water. I'm talking about rain, flood, drought, snow. Not water. Those have nothing to do with water. And when the forecaster says something about precipitation, that's not about water. Water is something you can put in a pipe or bathtub. You can drink it or wash dishes with it. It's real and managed.

But this other stuff—rain, snow, precipitation—that's ephemeral, out there in the metaphysical. I can't do anything about it. I can't change it, add to it, subtract from it. All I can do is complain about it, and by jingo I'm going to do plenty of that. After all, that's what I do about marriage, church, politics and other esoteric things in my life. This, dear friends, is the mentality of the average farmer.

In the sheer ecstasy of being a lunatic farmer, I have this notion that precipitation, rain, and snow have something to do with water. Since water, whether it's in a stream, rain cloud, or aquifer, is something visceral that I can see, handle, touch, and manage, I'm going to talk about water. The fact is that farmers can do a host of things to better manage water. But, like all of us, being a victim of circumstances (weather), absolves me of responsibility to take corrective action. It's a lot easier to complain and be a victim than fix it. Pardon me, but for the rest of this chapter, I want to talk about fixing it.

P. A. Yeomans, an Australian, wrote a book titled *Water for Every Farm*. In it, he put forth the two objectives of water management:

1. Every farmer should try to eliminate surface runoff from the land. We're not talking here about damming up rivers. We're talking about surface runoff during heavy precipitation events. Surface runoff is what swells streams and creates floods.

2. No farmer should end a drought with a pond full of water. Water is meant to be used, not just stockpiled for pleasure.

Yeomans went on to invent the keyline system and literally created an oasis in an arid land. Even in dry areas like Arizona or Utah enough downpours occur to significantly alter the landscape if all that water could be kept where it fell and dispensed to the land slowly. Think about the freshets and the incredible runoff that occurs during the winter and during spring rains. Keeping water where it falls, instead of letting it get away, is a far better savings plan than Certificates of Deposit in the bank.

Another man named Louis Bromfield, icon and pioneer of American sustainable agriculture in the 1940s and 1950s and author of several farming books including *Out of the Earth* and *Malabar Farm*, advocated ponds. He posited that the way to stop flooding on the Mississippi was not with expensive downstream Army Corps of Engineers projects on the mighty Mississippi. Rather, it was to create thousands, even millions, of farm ponds up in the headlands to catch that water before it built up volume and velocity.

The destructive force of water is always determined by volume and velocity. As it accumulates, going downhill, it builds up both to a deadly force. If the water is held where it initially falls it never builds up that destructive force. It stays spread out where it is needed and can be used, rather than accumulating in one major artery touching relatively few acres of land. From a hydrologic standpoint, ponds located high on the terrain create a seepage opportunity for everything downhill. Springs, streams, wells. Everything.

Back in the 1940s and 1950s when the old Soil Conservation Service began, it encouraged farmers to build ponds. SCS agents would design ponds and oversee their construction. The SCS even offered some cost share money. Up until recently, farm ponds have always been considered an asset. No longer. Now they are a liability.

In fact, government agencies are now using satellite photography to inventory farm ponds in order to identify these liabilities. You see, farm ponds attract water fowl. Water fowl, according to accredited government experts, are the primary vector for transmitting avian influenza. Never mind that studies conducted in Britain showed that poultry eating a bit of fresh grass every day were virtually immune to avian influenza. Never mind that the whole problem began in developing countries trying to mimic American industrial overcrowding techniques but without the modicum of sanitation practiced in American poultry houses, or Concentrated Animal Feeding Operations (CAFO).

Avian influenza, like all diseases, is encouraged by overcrowded and unsanitary conditions. This stresses the immune system, breaking through protective barriers and sickening the critter. The tentacles of the industrial food model extend far beyond the CAFO. Anyone who thinks the CAFO's assaults are limited to that community has no idea that the problems created by CAFOs reaches right into the remotest farm pond.

I love ponds. I don't ask for any government help to build them. Every time we scrounge together a little extra money, we call an excavator out and build another pond. On our farm, we've built more than a dozen and hope in the next few years to build another dozen. You can never have too many ponds because you can never store too much water and you can never be too diligent about keeping rain where it falls. Ponds do not reduce springs, creeks, streams, or rivers. What they do is even out the base flow of all these things to reduce flooding and turbidity.

Ponds are only loved by lunatics now that officialdom has decided to demonize them. Does it really make sense that the mallard landing on your pond threatens the world's food supply? Sounds like good science to me. Why stop with ponds? I think we should go ahead and exterminate all water fowl. Let's put a bounty on wood ducks, mallards, and geese. Wipe them out. Then we'll all be safe. Our taxes actually pay the salaries of people who think this is not satire, that it really would be a good idea. These folks are scary.

My problem with wells, which are the darling of cost share environmental programs as well as most irrigation projects, is three fold.

1. They poke a hole in the earth. We're talking about groundwater, about aquifers. This is a shared resource, and a precious one at that. But every time you poke a hole down into it, that's an additional opportunity for adulteration. A hole compromises the protective layer of sand and clay that water goes through on its way to the aquifer.

2. You can't look down and see how much water you have. Our farm is in what is called a carst geological formation. We have lots of caves, sinkholes and underground streams. The water percolates through the limestone and shale bedrock as it moves downhill. The water table fluctuates. We have a 1,000 foot mountain on our farm and right on top is a spring. How that happens is still a mystery, but it shows that water pressure occurs underground. Wells go dry without warning. And you never know when one will fail. A pond is like a visibly measurable storage tank. You can see what you have, and that's comforting.

3. It's everybody's water. An aquifer is shared by everybody. Just because water exists under my house doesn't mean it fell here or that somebody downstream isn't depending on it. I have the same thinking toward rivers and streams. That water did not originate here and it's not staying here. If all the effort to divert the Colorado River to irrigate golf courses around Phoenix had been used to eliminate runoff from the golf courses in heavy thundershowers, the Colorado could continue its unmolested journey to the sea.

When I catch water where it falls and hold it there, releasing it slowly, that sustains all the downstream water resources. In Australia every residence, both urban and rural, has a multi-thousand gallon roof collection water tank. In Colorado, it's illegal to have a rain barrel under a downspout. In California, you have to get a permit to build a bathtub-sized pond. I beg you environmentalists and anyone else encouraging this nonsense to understand that water impounded in small holdings way uphill eliminates flooding and sustains downhill hydrologic equilibrium. This is called 'base flow.'

If I have a water barrel under a roof downspout, that water is much more beneficial if I don't let it join the runoff from everyone else's roof at the same time. I might use that water to grow some carrots in a raised garden bed. That water then percolates into the soil and gradually goes into the groundwater and holds the water table up higher. If all of it must go downstream when the rain falls, it creates a flooding problem downstream and nothing is left to percolate slowly into the stream a month later. It's boom and bust.

On our farm, these ponds we've built gravity feed five miles of piped water. No pumps, no electricity, no energy. With the gravity, we have 80 pounds per square inch (most domestic systems operate at about 40 pounds) of pressure. It's like a fire hose. At intervals we have a valve that we can hook into for the livestock or even to irrigate. We don't have enough water impounded yet to irrigate significantly, but check back with me in ten years. It's coming.

Building a pond is not all that complicated, especially if you have a good track loader operator. We look for a swale and begin digging a bowl, using the dug out dirt to build the dam. I won't get bogged down now in the technical aspects of earthen pond building, except to say that in a couple of days and just a very few thousand dollars you can impound a million gallons of water. For much less than the price of most farm machines, you can impound a million gallons of water. For the price of an average tractor, you can build several of these ponds.

The point I'm making here is that building a pond, all things considered, is neither time nor capital intensive. When droughts come, we know we have millions of gallons of water stored up in high ponds that are as close as the nearest valve in our piping system. Gravity doesn't fail. If the power goes off, we have water. Over the years, I've watched farmers panic when drought comes. They liquidate cattle. They spend all day hauling water in little portable tanks that they fill by putting a pump down in the river. If everybody did that, the river would soon be dry.

The governor declares the county a drought disaster area to enable the farmers to receive low interest loans to help them through the crisis. And just because they've installed a state-of-the-art Natural Resources and Conservation Service (NRCS) designed system doesn't mean they are protected. Remember the previous chapter about the CREP project on one of our leased farms.

As I've watched this panic over and over during our years living here, I've come to realize just how little farmers think about correcting their water shortages. When the drought is over, they head to the equipment dealer for another tractor. They just don't think water is a problem that can be corrected. Just like they think soil can't be built. Just like they think the only way to feed the world is with Tyson confinement chicken houses.

The tragedy is that many of these farms have been in the same family for six generations. They are what we call Century Farms. Compared to most farmers in our area, we're newcomers because we've just been here fifty years. We have already built enough ponds to store enough water to weather any conceivable drought. Nor are we being presumptuous enough to stop. We're still building. Wouldn't you think that someone in that last century or more would have thought about storing some water and getting off the victimization treadmill? If they had just not purchased one tractor, or one truck, or one hay baler, they could have insulated themselves from one of the farmer's worst nightmares.

I believe one of our responsibilities as stewards of the land is to build more forgiveness into the landscape. Farmers should be shock absorbers. Nature can send some shocks, weather being perhaps the biggest one. But that is a given. We know a drought is coming. We know a flood is coming. It's our responsibility to bring cleverness and ingenuity to the landscape so that it's more resilient. Anything less is not good stewardship.

Here's a resilience vision for you. The Shenandoah Valley is roughly 80 miles long and 20 miles wide. If just half the soil moved to grow corn and grain to feed herbivores, which aren't supposed to eat it anyway had been moved to build ponds up in the mountains and high ground during the time the Europeans have been in this area (since 1740) by now we would have no floods and no droughts. Water pipes would run out of the mountains to the valley and we could have a veritable Eden, even in a drought. And it would all be done with gravity. No pumping. No energy.

Can you imagine? If you meditate on such a vision for a little bit, you begin to realize just how much energy, both human and petroleum, and creativity we've squandered...and continue to squander. And now that officialdom has demonized ponds, such a vision is even further from reality than it was a hundred years ago. We're not talking here about giant earthworks. This isn't about Hoover Dam. It's about little 20-foot dams scattered around the swales, like stairsteps, up and down the high ground.

In such a landscape, we could capture two or three times as much solar energy and turn it into carbon for decomposition. We could grow soil twice as fast. We could sequester twice as much carbon due to aggressive summer growing capabilities. We would make almost no hay because now a dry fall would be corrected with winter-stored water. The Potomac would not run high and muddy in the spring or after an August hurricane. Spring flow would even out. Stream flow would even out. Fish would be healthier. It would be an Eden.

This is what people are for. We're not for arguing about the pollution caused by a Tyson chicken house. We're not for going to McDonald's to buy a burger from a Colorado feedlot. We're not for filling out forms to give low interest loans to farmers who refuse to correct their water issues. People are for looking at problems and solving them in ways that work for everybody...for a long time. Not just a bandaid for today, but looking a century down the road. Instead, we've spent our energy on corn we shouldn't have grown to feed cows who shouldn't have eaten it to acidify the rumen to grow *E. coli* that shouldn't have lived to make people sick that should have been healthy, to fill hospitals we shouldn't have needed.

Once in awhile a pond leaks. We've had several do that and we use pigs to seal them up. Just sprinkle whole corn around the pond sides and put big pigs in. They have to be big in order to chew the hard corn. Pigs don't have to talk to lawyers or fill out tax papers. They can spend all their days figuring out how to make the ground impervious so it will hold water. Pigs do just about anything to make a wallow—a mudhole.

As the pigs eat the corn around the edge of the pond, they tread in the dirt and turn it to concrete. Pigs use their sides to screed across, just like a plasterer uses a trowel to smooth out the plaster. We've had excellent success using the pigs as a tamper and troweler to create an impermeable seal. We call it "Squealer

Sealers." I think we'll start a new business fixing leaky ponds. Just another example of the sheer ecstasy of being a lunatic farmer.

Hopefully a discussion like this helps all of us appreciate that farmers ultimately control, for better or worse, the ecology of the culture. It brings to light Allan Savory's sobering warnings that no single farmer can create an ecologically beneficial environment totally. We are all dependent on each other. If desertification happens all over my county, I can't stop it on my little piece of it. Ultimately, the water flows and transpires and moves among all of us. If you're wondering whether or not to purchase from a certain farmer, ask to see his ponds. If he looks at you like you have two heads, he might not be as environmentally friendly as you're lead to believe.

Just as I enjoy lying in bed thinking about all those dancing earthworms out there salivating on lignified carbon, I also lie in bed listening to rain, knowing that it is collecting in our ponds up on the mountain. As these ponds age and mellow, they become nesting sights for waterfowl, watering holes for deer, and even recreational splashing locations for bears. They encourage newts around the edges, salamanders, and toads. On summer nights, the croaking bullfrogs fill the night air.

One night a nearby farmer came over and we were out on the lawn talking. He was mesmerized by the croaking bullfrogs from the pond that provides water for our house garden. He had never heard them before. I thought how sad that a fellow would grow up surrounded by the abundance that's possible here, and never hear bullfrogs on his own farm. He had to travel to a neighbor's to hear them.

I've never been to a water park. I get all the thrills I need when I'm trying to fell a tree and it gets hung up in the crotch of another one. Then I have to go around and cut the supporting tree so both can come down together—right where I was standing to cut them down. Honestly, that's all the thrill I need. But when I see videos of the exuberant screams at water parks, I realize that I get just as big a thrill looking at happy cows and chickens in July drinking snowmelt water that provided habitat for a clutch of wood ducks and a croaking bullfrog. That is the sheer ecstasy of being a lunatic farmer.

TAKEAWAY POINTS

1. Regardless of what experts may say, ponds are assets, not liabilities.

2. Ponds offer more opportunities than wells.

3. Excavation devoted to building ponds is usually better than when it's devoted to growing corn.

Chapter 6

Toxin Free

In this section about nurturing the earth, we've talked about building soil, sequestering carbon, enjoying seasonality, local sourcing, and water retention. I want to end it by discussing, generally, the idea of do no harm. Or you could say keep it clean. Or no dumping. Or keep it toxin free.

The earth has a tremendous capacity to heal. Thankfully, every biological entity has that capacity. That's what differentiates the biological world from the mechanical world. If you're driving down the road and you hear a clunk, clunk from a sick wheel bearing, you can't pull the car off to the side of the road and rest it until the bearing feels better. You can let that car sleep, change the oil, fill the gas tank and flush the radiator, but when you get back in and start to drive, that bearing will be just as sick as it was when you pulled off. It doesn't heal.

Biological systems, however, do heal. Just like that cut on your arm that stops bleeding, then scabs over, then builds new flesh underneath, the earth has the same capacity. So to all my friends who listen to Fox News and conservative talk radio, I agree that the radical environmentalists are over paranoid. Oil spills on the beach clean up much faster just letting nature do its miraculous

66

cleaning than when people blast everything with steam jets. One volcano puts more noxious stuff into the atmosphere than all the coal-burning power plants combined for a long time.

The capacity to heal is remarkable. But just because the earth can heal does not give us license to disregard its balances or abuse it. I think too often the conservatives dismiss toxicity on a whole planet scale, believing the earth is too big to destroy. That may be true. But while the whole planet is experiencing lots of pin pricks, if I live on a pin prick, that's a big deal. We can destroy a little piece without jeopardizing the whole planet.

Wendell Berry eloquently points out that before we begin discussing how to live correctly on the whole planet, it seems reasonable to figure out first how to live correctly on a little piece of it. And that is where I want to go with this discussion. Generally, I think planetary health is way too big to describe, grasp, or fix anyway. The old Chinese saying that if everyone would sweep in front of his own doorstep, the whole world would be clean seems appropriate here. The question is not what those yahoos over there are doing, but what am I doing?

We've already talked about depriving the land of carbon when we sell hay. Some farmers pride themselves in building carbon by only buying hay. But in the big picture, that is translocating carbon. Here at Polyface, some accuse us of doing the same thing since we're buying grain for pigs and chickens. Confession: the pastured poultry, both meat and eggs, are the least sustainable portion of our farm. There, it's out. Historically, poultry primarily and hogs secondarily have been scavengers. Poultry especially was a luxury. When President Harry Truman expressed his American vision as a chicken in every pot, it was because chicken was Sunday dinner. The special luxury.

Not until the Transcontinental railroad, combines, petroleum, antibiotics, confinement housing, and automatic feather pickers became widespread, could large scale poultry compete with herbivores. Herbivores were always Everyman's food— grazing on perennials. Nothing had to be plowed or fertilized. A carcass is much easier to access when it's covered in a hide than in feathers. The labor required to access a pound of beef is way less than what is required to access a pound of chicken—historically.

Chickens cleaned up table scraps and gleaned around the backyard and barnyard. My wife Teresa's grandmother said that as a little child growing up in the early 1900s if they wanted chicken for the Fourth of July, a hen had to begin setting on eggs January 1. The chicks would hatch by the end of the month and then it would take five months for the cockerels to get big enough to butcher. Today, with modern genetics and all-you-want feeding, that five months has been compressed to two.

So we could argue that even the thought of commercial poultry flocks is a bit unnatural. People should keep a few chickens in their backyards. Pat Foreman has written an excellent book titled *CITY CHICKS* describing the ins and outs, as well as the very real landfill reduction reasons, to have chickens in the backyard. If everyone who could do it would just do it, the industrial poultry industry in America would not exist. And chickens would reduce the waste stream the way they've done for millennia.

We're not there yet. People want to buy chicken. So at Polyface we grow chickens. And we buy genetically modified organism (GMO) free grain from neighbors. Can grain be grown and transported off the farm ecologically? Yes, but with the following constraints:

1. Rotation with pasture. The historic seven year rotation, in which corn was followed with small grain followed by a legume and then four years of grass is a working model to maintain fertility. This still works in Argentina, where fertilizer is too expensive. The organic-matter building years of legumes and grass between the two tilled years holds the fertility to reasonable levels.

But could we grow all the grain we need if everyone went to such a lengthy rotation? Yes, if we quit feeding herbivores grain. The lion's share of the grain grown in the world goes through herbivores. A relatively small portion goes to people, pigs, and poultry. If we terminated grain feeding—including corn silage—to herbivores, it would significantly reduce the amount of grain currently sought on the world market.

2. Returning all leaves and stems to the field. The whole rotation breaks down, of course, if during the grass years we extract all the grass in the form of hay and sell it off farm to

someone else. The whole rotation scheme assumes that the only thing leaving the field is kernels of grain. Straw, stalks, grass, legumes—everything must be recycled on location. That doesn't mean it has to be fed there. It could be fed in a nearby shed or barn, but the nutrients from that carbon must be returned to the field whence it originated.

Old-timers around our area tell me that one of the earliest chores for pre-teen boys was to go out in the winter and pick up cow pats dropped in the barnyard overnight. Forking these cow pats into a wheelbarrow, the boys would take them into the barn, under roof, and spread straw on them to make sure they didn't leach away in the next heavy rain or snow. Now that's being careful. That bedded dung, then, can be retuned to the field whence the hay or grain came.

3. Long stemmed grain varieties. Plant breeding for nearly the last century has centered on growing a bigger kernel on a smaller stalk. I don't have a problem with the bigger kernel part, but why a shorter stalk—the part that becomes straw? Harvest efficiency. A combine cuts off the plant and sends everything through a set of screens, shakers, and fans to separate the seeds (grain) from the husk, leaves, and stalk. The grain goes in a storage bin and the straw dumps out on the ground behind the machine. The less stalk and leaves the combine has to shake through, the faster the machine can go. Faster means more ground covered in a day, more grain harvested in a day, more efficiency.

Of course if the grain is going to exit the farm and the straw is going to stay, a more sustainable fertility cycle occurs when the amount of staying carbon is as high as it can be—to feed the soil to compensate for the exported carbon (grain). Historically, straw was considered an asset because it was the bedding of choice for horses. As the need for horse bedding diminished, so did the market for straw. Straw took a double whammy: the market diminished and combines didn't like straw.

4. Flat ground. America's Midwest is certainly more forgiving for grain production than hilly ground. But lots of flat ground exists in our mid-Atlantic region. If the ground to be tilled is in small pieces and judiciously picked, grain can certainly be grown in an earth friendly way. Since mules did not fall over on hillsides, historically even very steep hillsides were plowed and put

into grain production. But gradually they eroded and farmers were forced onto gentler ground.

If grain were limited only to the land conducive to tillage, and all of the previous protocols were followed, grain could be part of a healing regimen. Violating any of these principles compromises the ability to grow grain in an ecologically responsible way.

With all that said, let's return to the grain we buy at Polyface to raise chickens, agreeing that in the perfect world we probably would not raise as many chickens as we do. Certainly in principle, grain can be grown and moved about without jeopardizing localized fertility cycles.

Now that we've dealt with the grain issue, let's turn our attention to the birds and their manure. Even though the birds are on pasture, their manure load is significant. If we just consider nitrogen, as far as the soil is concerned, only a certain amount of nutrient can be ecologically metabolized.

Too much nitrogen is too much nitrogen. Whether it's organic or not doesn't matter. Too much of anything can overload a system and develop toxicity. Or as it flushes through the system, it dumps pollutants into the ecosystem. What we want is just enough nitrogen to keep things growing well, but not enough to overload the field's capacity to metabolize the nutrient into plants.

Because this has become such a big issue in the industry, agronomists have actually created Best Management Practice (BMP) recommendations for nutrient loads in different climatic and cropping conditions. In our area, pasture is rated for about 120-150 pounds of nitrogen per acre per year. That is assuming normal pasture conditions, which include continuous grazing. Since that's what everybody but lunatics practice, the limits assume continuous usage. Obviously, with good rotational grazing, we can double the forage production, which in turn can double the nitrogen metabolism.

In our portable pasture broiler shelters, which are 10 feet X 12 feet by 2 feet high, we put in 75 chicks at 2-3 weeks of age and run them up to 8 or 9 weeks old before processing. By moving them every day and keeping them controlled and sheltered in those floorless schooners, we get an even manure distribution. Unlike a more free range or day range concept which concentrates manure in shelter, feeding, and watering locations, our birds fairly evenly cover the ground.

The total square footage covered by every 500 birds in their lifetime is about 1 acre. And they are putting down about 200 pounds of nitrogen, which is right in the ballpark for metabolic activity. Interestingly, if you calculate what each of those birds eats during that period of time, it takes exactly one acre to grow the grain. People ask me how I know that this level of chicken density is right.

My sense is that if the manure load is at the limit of soil metabolism and the land parallels the grain requirement, that's a confluence that gives credibility to the chicken population density. What this means, of course, is that we try not to cover any area twice in a year. We want essentially a year of rest to allow several grass growth cycles to metabolize all the nutrients before returning with another application. When we spread litter from the brooder house or winter laying chicken hoophouses, we put that on ground where we have not run broilers.

Compost from the cows can go on ground where the broilers ran. The compost is loaded with undigested pieces of hay, wood chips, and sawdust which give more carbon for the nitrogen to break down. This symbiosis between different types of manures from different types of animals at different times of the year is what creates ecological harmony. That's real harmonic convergence.

Now that we've gone through the math and thought process, we all need to pause a minute to appreciate that the industry does not think this way. When a farmer decides to locate a confinement chicken house on his property, nobody asks him if his farm can metabolize all the nutrients. The poultry company, the farmer, the lending institution, the building inspectors, and the construction crew all follow the universal assumption that the manure will be fed or spread somewhere else. These folks don't even think about a relatively closed production model. It's a factory—raw inputs in the front door and packages out the back door. It's all linear rather than cyclical.

At Polyface, all the manure cycles back on the farm where the birds live. The birds are processed on farm and their guts, feathers, and blood composted on site. Only the carcasses leave the farm. In a perfect world, we would grow all the grain for the birds and run them on the grass areas between the grain rotations. But our farm has been plowed way too long and doesn't have grain growing land. So here we are: not perfect, but at least thinking and trying to have as localized a cycle as possible.

As a result, we have identified a carrying capacity for our farm. If we have more market than our carrying capacity, we either need to say no to that market, or figure out a way to expand that maintains this carrying capacity integrity. And that is what we're doing with leased farms, subcontractors, and former apprentices or interns. We'll talk more about that in a subsequent chapter, but for now be assured that's a far cry from just laying on more poultry here and creating a toxic nitrogen dump.

The main point I wanted to make by going into this level of detail is the number of things we bring to the equation when we talk about ecological balance on our farm. None of this even enters the minds of industrial practitioners. It's all irrelevant poppy-cock. But what a joy to be able to sleep at night thinking of all these symbiotic circles, all this perfect humming, balanced ecology. That's sheer ecstasy.

When you realize that this kind of thinking is not occurring on an individual farm level, it ramps up to a new level of incongruity when you add up all the farms involved. Soon it becomes a whole bioregion. Suddenly you have the Shenandoah Valley creating a toxic nutrient situation and having to export manure a hundred miles to get it far enough away. Or feeding it to cows. Or generating biodiesel from it. Why not just have the grain and chickens grown near each other so the whole thing can be environmentally and economically symbiotic? Why do you have to be a lunatic to even ask that question?

I can't help but share another manure-related story that dates back many years to when the Chesapeake Bay Foundation pushed through government funding to stop non-point source pollution. The culprit? Manure. Primarily from feedlots and confinement dairy operations. The answer? Taxpayer-purchased manure lagoons.

Notice the sequence of thinking. First, we take cows off of pasture. We begin mechanically growing and harvesting grain and forage for the newly-confined cows. In order to keep them from being in mud and manure all the time, we have to pour concrete, bend rebar, and build confinement facilities. Now we're emotionally and economically locked into extremely capital intensive infrastructure.

We can't let the manure build up on the concrete because that's yucky and unsanitary for the cows. So now instead of the cows dropping widely spaced manure around the pasture, it has to be mechanically scraped or flushed every day. According to government nutrient management specialists, whose paradigms assume water-based human sewage systems, the best way to handle manure is in water. This concrete-based system involves no carbon, so composting isn't an option. As a result, excrement must be handled as a slurry. We can't spread it every day because lots of days aren't conducive to spreading—like when the ground is frozen or snow covered or muddy. So we have to store the slurry.

That's where the taxpayers come in, because slurry storage systems, whether earthen lagoons, metal lagoons in blue metal tanks, or concrete lagoons like in-ground swimming pools, are expensive to build. The poor, destitute farmers need some free money. So vegans, vegetarians and foodies get to pay for manure lagoons at the government's behest. What charity.

These lagoons are anaerobic and toxic as sin. The acidified manure chews away at the metal and concrete lagoon, giving these structures only a limited life. This slurry cannot be handled by front end loaders that almost every farmer has. Instead, the farmer can buy specialty slurry pumps, augers, and spreading equipment—which acidified slurry chews up voraciously. And to top it all off? The acidified slurry, when spread on the field, kills earthworms on contact and needs—you guessed it—calcium fertilizer to correct the acid.

If you attend any conventional farm show in any state, 90 percent of the equipment, infrastructure and tools on display are to fix problems that we would never have had we been thinking about doing no harm in the first place. The day farmers decided that a farm is a factory and not a cycle is the day this whole plunge into nonsense occurred. Without appreciating the earth's balance in the soil, the beautiful nutrient cycle between perennials and herbivores, and a solar driven system, farmers moved cows off the pasture and created a whole sequence of abuses.

The feed now had to be mechanically harvested and toted to the cows. Concrete had to be poured and payed for. Cow feet became sore because concrete isn't spongy like pasture. Manure had to be daily scraped to an expensive lagoon. The anaerobic manure lagoon became more acidified and toxic. The soil itself suffered insult as the final victim in this sordid chain of events.

As the world turns green, we have farmers taking the next step—often at taxpayers' expense. Now they are stretching balloons over the lagoons and receiving rewards from environmental organizations for capturing the methane. Folks, this is like giving a Compassion award to a doctor who goes out and breaks children's legs just so he can fix them with compassion.

Who dares to ask: why do we have the cows on concrete to feed a lagoon that generates methane in the first place? The farm wouldn't need 80 percent of its energy if the cows hadn't been confined on concrete in the first place.

Now, for sake of discussion, let's assume that in certain periods of the year, especially in northern climates, we need to put the cows inside for a couple of months a year. At least we're not dealing with the health and nutrient problem 365 days. But if we're going to put them in, like we do here at Polyface for maybe two months, we'll use carbon as a bedding. Then the cows have a nice soft bedding pack. We can make compost and spread it with cheap manure spreaders using the front end loader that every farm already has. And the manure is healthful for the soil, not toxic. It feeds earthworms and smells beautiful. But that's not on the land grant Best Management Plan. So it's lunacy.

I've gone into some detail on the manure issue, because ultimately everything runs on manure. Not really, it just seems that way. But I've used manure as the focal point of this discussion because it represents the most egregious problem in industrial thinking. During the 1970s agronomists all over America exhorted their farmer disciples that manure was worthless. They said it didn't even pay to spread it. And everybody laughed farmers like us to scorn for composting it, handling it, spreading it and treating it like royalty.

Fast forward two decades and now the same bureaucrats are telling farmers to care for their manure, have it tested, apply it with dignity and try to keep from buying so much expensive petroleum-based fertilizer. You want to know a dirty little secret? I was right all along. Those guys told a lie and then tried to correct it. I never lied.

I could belabor the earth toxicity discussion with many more examples. Fumigating the soil in commercial strawberry production in California is a good example. Let's just list some alternatives:

1. Don't grow as many strawberries in California.
2. Don't demand strawberries year round—anywhere. Eat seasonally.
3. Use more compost to balance the soil.
4. Make the fields smaller and/or interplant.
5. Reduce the genetic hype so the plants are hardier.
6. Use foliar sprayers based on seaweed or manure tea.

I'm not a commercial strawberry grower, but from folks I've talked to who don't fumigate the soil, plenty of alternatives exist. Lest you think I'm being unfair by saying things like "don't grow as many strawberries in California," be assured that for decades Wall Street type capitalists have come by our farm encouraging me to be the Don Tyson of pastured poultry. The temptation to just grow more birds here, close to our house, is huge. And some growers are doing that.

But down the road, failure to recognize the carrying capacity, and failure to adhere to the toxin-free rule will come back to haunt us. We've become a nation of technicians enamored of the how but not the why. It's time to realize that the most important things in life are not on balance sheets and tax statements.

When is the last time you presented a business plan to a bank loan officer and had her ask: "That's all well and good, but what does this plan do to salamanders? What does it do to your marriage? What does it do to your kids? What does it do to your neighbors? What does it do to the earthworms?"

Unless and until we begin asking those kinds of questions, we will continue assaulting nature. That, in turn, will require more sophisticated technology, more infrastructure, more questionable concoctions from pharmaceuticals to pesticides, more energy, and more problems. Most of the problems we research are a result of not keeping things toxin free. Yes, I know anything can be a toxin. You can't just drink a bottle of organic foliar spray without getting sick. Manure is a toxin.

The point is to handle things so they aren't toxic. Respect the cycles and natural cleansing that rotations, decomposition, and sustainable carrying capacity affords. If we devote ourselves first to that, instead of solving the problem created by not thinking about that first, we will end up much happier and healthier. Indeed, we can enjoy the sheer ecstasy of being a lunatic farmer.

TAKEAWAY POINTS

1. Every ecosystem has a carrying capaicty.

2. Most of our problems were created by us with incorrect thinking.

3. Manure is best handled by composting.

Produce Food and Fiber

Chapter 7

Growing Stuff to Eat

D on't all farmers grow stuff to eat? That seems like such a silly question. But in actuality, eating quality doesn't register on most farmers' radar. The fact that this stuff gets eaten takes a back seat to packaging and shipping.

Commodity agriculture is fundamentally concerned about one question: does it fit our box? Every item has a box, and if you're outside that box, steep price discounts are yours to enjoy. In beef cattle, black is the color of the box. Any other color doesn't attract buyers as much.

In dairy, black and white is the color of the box. Nothing else is as important. This myopia drives people like the American Livestock Breeds Conservancy crazy. ALBC is devoted to preserving the genetic diversity that naturally exists among minor breeds—all those breeds for which the industry has no room.

To be fair, the industry response to this is that it has done the feeding trials and research and market studies and the box they've picked is the best box. It's the most efficient box. It's the most salable box. And while that may be true for what they were looking for in their research, who decided the most important

78

characteristics to look for? In the beef cattle industry, for example, as grain feeding and feedlots became more widely used during the 1960s, the industry needed cattle taller to handle the manure and conditions in feedyards.

Obviously, this had nothing to do with food. I have a friend who tells the story about going to a beef cattle convention and listening to Ph.D. after Ph.D. talk about how to grow animals faster, bigger, cheaper. Finally, toward the end of the convention, he asked the professor: "How can I make the meat taste better?"

He says the expert wrinkled his brow and responded with another question: "You mean how do we grow them bigger?"

My friend: "No, sir. I mean make the meat better?"

To which the expert queried: "Why would you want to do that?" You see, the fact that this was food never crossed the professor's radar.

Tall cattle don't make better meat. All that air underneath their bellies doesn't add taste or nutrition to the plate. And the height makes them much more difficult to finish on grass. Those of us producing grass finished beef want little barrels on toothpicks. What I call state of the art 1950s genetics. And we want smaller phenotypes. We don't want behemoths that can't ingest enough forage to keep their boiler stoked. A lot of these animals have to eat corn in order to perform. Their mouths and bellies just aren't big enough to do it all on forage. But that's the box.

In grass finishing operations, you'll see all kinds of colors: white, gray, red, brown, mottled and even blue. When you go to a steakhouse that advertises Angus beef, that's a black box. I have no vendetta against Angus. Those guys have been incredibly successful at marketing their mystique. But eating quality depends on a lot more things than the color of the hide.

One of my favorite things to eat is an apple. It doesn't need to be cooked, sliced, diced, pureed, sautéed or anything. Nothing beats the sheer ecstasy of biting into a crisp, juicy, sweet apple. Here's my question: have any of those Washington State red delicious commercial orchardists ever actually eaten one of their apples? Come on, be honest. They're soft, tasteless, pithy mish-mash. That's not an apple. I don't know what it is, but it's not an apple. It might be red. And it might grow on a tree. And it might have a stem. But it's not an apple. It's an abomination.

No wonder most food now is processed rather than eaten raw. No wonder the produce section gets short shrift in the supermarkets. The real money is in doctored stuff. It's breaded, pre-cooked, seasoned, food colored and texturized. That's because farmers aren't growing stuff to eat. If they were, you could walk through their farms and eat things. And people would enjoy the raw stuff.

One of my favorite white tablecloth dining experiences was at food maven Alice Waters' Chez Pannise restaurant in Berkley, California. I don't remember what I had for the main course, but for dessert she brought a bowl of clementines from Michael Ableman's farm. The explosion of taste as I bit into that first wedge still makes my mouth water just writing about it. Excuse me while I wipe the drool off. It was the stuff of legends. Who would think that a raw fruit could be a dessert item in a white tablecloth restaurant? I thought dessert had to be some swirly chocolate mousse. Now don't get me wrong—I love confectionaries, pastries. Oh yeah, that's good stuff.

I don't think I had ever been served something unprocessed for desert. But so confident was Alice in the eating quality of Michael's Clementine that she put three of them in a bowl, garnished them with some California almonds, and served them just like that. Straight from the earth. Straight from the farm. A farm that produces stuff to eat. No doctoring. No embellishment.

How about tomatoes? Now there's a beautiful critter. Why are tomatoes the most commonly grown backyard food item? Probably because they're the hardest to ship and keep palatable. But ship they do. Tractor trailer loads of them. From 3,000 miles away. Be honest, now. Almost everyone has eaten a backyard tomato. You know the kind. When you slice into it juice runs down the knife and pools up on the cutting board. The structural webbing inside glistens iridescently, standing out from the delicate seed-juice innards. Do you know what that tomato would look like if you put it on a tractor trailer for a week and sent it 3,000 miles? It would be a flattened glob of pulp.

I ask the question again: have those tomato growers eaten their own tomatoes? They are like cardboard. Hard, tasteless.

Yuck. Who wants it? You would never eat that tomato as a sliced tomato. You might dice it up and put it on a Taco. You might cube it and add it to a veggie-dip tray. But just to enjoy eating sliced tomato? Not on your life. That's reserved for those tomatoes grown close to your table, either by you or the farmer you love.

Chicken. Numerous chicken farmers won't eat the chicken they grow. Back behind the barn, out of sight, is their secret chicken. There the chickens receive scratch feed and kitchen scraps and run around a little. That's what the farmer eats. In fact, if you attend an industry livestock or poultry symposium, the speakers will all refer to what they do as the protein business. They're not even in the livestock business anymore. They are just making protein, to be fabricated, amalgamated, irradiated, adulterated, reconstituted and prostituted.

When Daniel was in 4-H attending banquets he got quite a kick out of telling the girls sitting at his table: "A chicken doesn't have this part. Did you know that this doesn't come off a chicken?"

And of course the grossed out girls would ask: "Well it's chicken, isn't it?"

Daniel: "But if you butcher a chicken, this piece of meat is not on it. It's not breast, not thigh, not leg, not wing. It doesn't exist."

Girls: "Ewwwww. Where does it come from?"

Daniel: "They glue bits and pieces together, run them through an extruder, and press it all together like plywood."

By this time, of course, the girls were all agape and making those pretty pout-type expressions as Daniel gleefully exposed the atrocities of the poultry industry.

Recently Rachel (our daughter) went to Washington D.C. with a former schoolmate who was visiting from the west coast. As they discussed the excursion the evening before, Rachel noted that the mall didn't have any place to eat.

Rachel's friend: "Oh, that can't be true."

Rachel: "No, really. You can't find anything to eat on the mall."

Rachel's friend: "Surely there's a McDonald's there."

Rachel: "Like I said. You can't find anything to eat."

If many restaurants aren't making anything to eat, just imagine how few farmers are growing anything to eat. I talked with a fellow who had just finished an animal science degree at a major land grand university. I made a disparaging remark about cattle growth hormones that almost all commercial growers use.

His response: "Well, consumers have been screwing the farmer for a long time and it's time for us to screw them back."

Well, that's a great spirit for a new college graduate wannabe farmer. Farmers by and large view themselves as just growing stuff. Just stuff. The fact that it's food that should give nourishment, taste, and texture doesn't enter the thought process.

We had a bunch of extra eggs one spring. Most of our pastured eggs go to white tablecloth restaurants and informed eaters. But with eggs running out our ears, I decided to try peddling them—at the going commercial wholesale price—to some local Staunton restaurants that I knew did a spirited breakfast trade. The first place said they were too big. I couldn't believe it. What do you mean, too big. Won't you taste one? Can't you tell your prep staff to use one or two fewer per dozen in your recipes? Crazy. Too big my foot. They just didn't want to fool with a separate vendor.

Next place: "our customers don't want eggs with taste. These have a taste. Our customers don't like that." Let this be a lesson to you: never try to sell real food to a cook. Always look for a chef. Again, I was incredulous. He just pronounced that as if it were an axiom: "our customers don't want eggs with taste." Have you asked them? Would you be willing to try?

I had a lady in our farm store recently who had a picky eater for a son. He was a diminutive six-year old with an attitude. And he didn't like food. Mom was desperate for him to find something he liked. She finally ended up on our doorstep. I encouraged her to buy some eggs and try them. A couple of weeks later she was back. "My son wants six eggs every morning. He's devouring them. I've never seen him eat anything like that." At Polyface, we're growing eggs to eat. Our first thought is about nourishment.

Pastured eggs contain the proper balance of Omega 3 and Omega 6 fatty acids. That balance is the key to cholesterol. Teresa's 90-year old grandmother had perpetual cholesterol problems, according to her doctor. One spring we hard boiled a bunch of eggs and pickled them for her, for a treat. Her doctor had forbidden her to eat eggs. Teresa and I knew that our eggs would be good for her, so we told her to cheat. Grandma loved pickled eggs. Well, she decided to cheat, and ate one. It tasted so good, she ate another. Before she knew it, she'd eaten that whole dozen in two days. The next day was her regularly scheduled doctor checkup that she'd forgotten about. Oh, she was in a dither, praying that the doctor wouldn't order bloodwork.

He did. She waited for the awful news. He came in beaming: "Your cholesterol is normal for the first time in 10 years. Wonderful." From that day on until she finally passed away at the ripe age of 100, she ate our eggs with gusto. This story was repeated about the same time with a National Guard officer who was fighting cholesterol. He began eating several pastured eggs a day: end of problem.

Back to the extra eggs at restaurants. Third restaurant: "these are brown eggs. I will never serve a brown egg in my establishment." By this time, I knew I was a certified lunatic trying to sell real food into the industrial food system. Folks, all these restaurants are still in business today, thriving. They are featured in food publications. They have a stellar reputation. I have news for you: they're serving junk. They're not serving stuff to eat.

This next story doesn't involve eggs. It involves ham. We have a local German restaurant with a good reputation. I took fresh pork ham over there trying to market it for schnitzel. The owner tried it. The cook tried it. Both of them remarked how pink colored it was. In case you don't know, when the pork industry positions itself in the market as "the other white meat," what it really means is "the other anemic slovenly tasteless junk meat." Hogs that gambole in the grass and forests develop muscle tone and nutrient density that expresses itself in color. Just like the best flowers are the ones with the most vibrant color. Cheeks of healthy children: color.

Anyway, the owner and cook were discussing the rose color and how different that was. By the time they put it in the batter and cooked it, all you tasted was the batter and seasonings, not the pork. It was covered up pretty well. Our pork, being real food, was a little more expensive than what they were buying. Since the taste test yielded nothing spectacular, the owner looked at me and said: "Well, our customers just want slop. That's all they care about. Just slop." End of conversation.

I wonder if she offered local Pigaerator Pork with a story if people would gladly pay an extra 50 cents. Goodness, she could offer both kinds. But see, if your heart isn't in it, how can you insure integrity anyway? She'd probably use the junk and call it ours.

Compare that to one of our chefs in a fine dining establishment the first time I took him our eggs. I explained to him that our eggs would fluctuate throughout the season because in the winter the layers couldn't get fresh grass. As I'm writing this sentence today, we have 23 inches of snow on the ground, and that's not conducive to pasturing chickens. Before I could even finish, he interrupted me: "Oh, no problem. In chef's school in Switzerland (all these famous chefs have some exotic mentoring or cheffing experience) we had recipes for March eggs and special recipes for June eggs and others for October eggs to accentuate the nuances of the egg as it adjusted to the hen's seasonal dietary changes."

Can you imagine? What a great appreciation for the splendor, the grandeur, the sheer ecstasy of food. Imagine changing the menu to accentuate seasonal nuances. This chef had priorities like mine. Packaging and shipping don't even enter into my lexicon. I don't even think about those things. I'm thinking about meals I'm thinking about eating. And I want to build our farm business around the nucleus of people in our foodshed who also think about eating.

Back when we used to sell at the local farmers' market, one day a blue-haired country club sophisticate came by our stand with her nose in the air and snorted: "$3 a dozen for eggs! I would never (actually, it was the Virginia sophisticated "nevah") pay that much for eggs." She happened to be holding and sipping a Diet cola.

I leveled my eyes right at her and said: "Ma'am (you still have to defer respectfully to these blue hairs since after all they do own the banks in town), there're more nutrition in one of my eggs than in a tractor trailer load of those 75-cent cans of soda you're drinking."

She harrumphed on down the walkway, towing her little leashed Fifi along at her heels. The dog probably ate off of china and lounged on a heated blanket. Anyway, I don't lose sleep over these kinds of people. Forget them. They aren't interested in stuff to eat either. They are just interested in ingesting.

Ingesting is really different than eating. Eating conjures up things like appreciation, culinary skills, chewing, salivating, conversing, pleasant aromas, family and friends. Eating is a social activity. All cultures are defined primarily by religion, architecture, and food. Eating is intricately tied to cultural identity. When you think of American food culture, what do you think of? Ingesting.

Ingesting is what Marines do at boot camp when they have 10 minutes to inhale processed material. Ingesting is what people do when they pop boxes in the microwave without talking to anybody and consume it while watching TV. Ingesting is what most people do because they don't want to take time to eat. The preparation, thoughtfulness, acquisition effort, and menu planning that eating require are meaningless to them.

I know some people think I'm being awfully hard on farmers, but I see this whole issue as being a bit like which came first, the chicken or the egg? As a farmer, it's easy to point my finger at modern Americans and dismiss my uncaring attitude about what I produce because after all, Americans don't care about food anyway. But turn the finger around. If farmers really produced stuff to eat, would Americans perk up and rediscover something special about food?

Certainly Chef Ann Cooper and others involved in the Farm to School movement have found dynamic acceptance among grade schoolers when they are exposed to real food. If you wait until college, the acceptance drops dramatically. I know on our farm we receive countless stories about children who won't eat store bought eggs once they are accustomed to the pastured variety. My own grandchildren will not eat commercial chicken or eggs. You can try to fool them, and they can tell it every time. "What's wrong with these eggs?" is a common question children ask in the homes of our customers when they've inadvertently run out of Polyface eggs and resorted to commercial brands.

Are farmers supposed to wait until all Americans embrace food again? And start eating instead of ingesting? Who raises the standard first? Who blinks first? I believe if farmers really devoted themselves to producing stuff to eat, Americans would eat it. I don't know if the processed food frenzy is all because of hectic lifestyles, lethargy, negligence, and convenience. What can be more convenient than eating an apple? Or a Clementine?

The last time I visited Eliot Coleman, gardening guru extraordinaire and author of arguably the best gardening books of all time, he was giddy over the fact that he'd just learned that his vegetables commanded the highest trading value among school children in his community. In other words, when children opened their lunch boxes and began peering into what was available in other boxes, Eliot's vegetables were the most sought after item at the table. His baby carrots traded higher than Little Debbie pastries. How about that?

It's because first and foremost, he produces stuff to eat. Not stuff to fill a truck. Not stuff to fit a box. Not stuff to process into indistinguishable microwavable mush. Just good stuff to eat. I submit that if more farmers just grew stuff to eat, and impressed eating on their frontal lobes instead of the logo from agribusiness industries displayed on their farm caps, the country would be a better place.

What a hoot to know that I'm producing stuff to eat. I walk around with visions of families gathered around the dining room table chomping, chewing, masticating, salivating, food fighting. Just enjoying eating. Together. Nourishing, delectable, memorable. Wow, the sheer ecstasy of being a lunatic farmer.

TAKEAWAY POINTS

1. Good food needs little culinary doctoring.

2. The food industry creates arbitrary objectives that do not include nourishment or taste.

3. Eating and ingesting are two different activities.

Chapter 8

Land Exercise

Just like people need exercise to stay trim and fit, the land does too. To be sure, rest is also good. But rest without exercise makes ecological couch potatoes. I introduced this topic in a previous chapter so I won't go through all the principles of freshness and disturbance here. What I want to do now is to broaden the discussion to include urbanites and especially land preservationists.

To review, we established that human activity can be healing or hurtful. Nothing about human presence on the land is necessarily good or bad. Bad activities create ecological sickness and good activities create ecological health. A growing number of people in our culture, divorced from farming, divorced from the land, believe that human presence on the land is inherently harmful.

I encounter this attitude routinely with the preservationist portion of environmentalists. I certainly don't have a problem being called an environmentalist. After all, I've chosen the moniker "Christian Libertarian Environmentalist Capitalist Lunatic." But environmentalism comes in different shapes and sizes. For example, Permaculture, which was invented by Bill Mollison and Dave Holmgren, is dedicated to designing ecologies that operate with as little outside intervention as possible. I've visited numerous permaculture sites, and they are unquestionably the zenith of environmental enhancement.

Only one problem. They require lots of landscape manipulation. From ponds to free form construction to water sluices to gray water biological filtration systems to mouldering toilets, the whole objective is to move things around and build infrastructure that operates on its own. And permaculture is about exercising the landscape with diversity to create more intricate synergistic relationships. In a swamp or wetland, for example, if you create a pond, you actually multiply the environmental zones in which plants and animals can live.

Environmentalist-sponsored government wetland regulations, however, do not allow that kind of landscape change because it would alter it from its present state. In deep permaculture ecology such a landscape change would attract a host of biological diversity because the new pond growth zones create new habitats. Instead of being limited to wetland critters, both plants and animals, the habitat can now support fish, offer a water heat sink that evens out temperature highs and lows, creating a unique micro-climate, and even offers a place to collect silt or attract waterfowl.

I am all about preserving greenspace; the question is how to do it. Creating large lot sizes and locking down the land is not the answer. I was recently in the Santa Barbara area of California and met with a rancher who can't even build a house on his family ranch because the preservationists have created a 300-acre one-dwelling per parcel stipulation. Heralded as land preservation, this regulation denies the second generation a domicile on the family ranch and threatens to destroy the financial viability of the ranch.

If the son can't live there, he has to commute from town. Now he's not living cheaply on the ranch. Suddenly, what could have been a viable agricultural operation is no longer viable. And herein lies the great fallacy of the preservationists: they aren't primarily interested in keeping farmers viable. What they want to do is freeze the landscape.

I can't help but chuckle as I drive through our county and see the proud signs on farmgates: "This land protected forever by conservation easement." Forever is a long time. Can you imagine some farmer on the outskirts of Rome challenging Attila the Hun: "Sorry, sir, you can't come across my land. See that sign there? It says protected forever. If your army comes across this farm, they will leave horse tracks. And I really don't like the way you picked the neighbor's flowers yesterday. Mine are protected forever. And if you don't leave, I'll call my easement trustees and have them deal with you."

This notion that we can just preserve open spaces by fiat is symptomatic of two things:

1.   A complete disconnect with the land and actually making a living from it. Most of the farmers who sign their lands up under these protection programs do not make a living farming. They have other sources of income. And the trustees who enforce the preservation easements, for the most part are urbanites whose agricultural perceptions have been formed by urban environmentalists.

2.   Hatred of private property rights. America's urbanization has fundamentally altered the historic mystique of property ownership. Land as possession has been replaced by land as public park. Many view farmland as a public recreation place.

Farmland represents the sum and substance of the average farmer's wealth. The farmer works that land, pays the taxes, keeps the buildings up, derives a livelihood from it, invests in it. It is the farmer's retirement account, investment portfolio, savings. It represents his life's total monetary value. At public hearings, when nonfarmers sit around a table and pontificate on what should or should not be done with farmland, I would like to ask a question: "Who wants to view your stock portfolio, your certificates of deposit, or your bank accounts in the same way?"

It's pretty arrogant and disingenuous for a bunch of nonfarmers with their life's work protected in banks, money markets, precious metals and stock portfolios to tell a bunch of farmers what they can and can't do with their life's material wealth. And don't start in with "farmland lost is farmland lost forever." Again, forever is a long time.

I was in St. Louis recently and visited several urban farms. Many people don't realize that from 1950 to 2000 St. Louis lost 50,000 people per decade. It went from one of America's 10 largest cities to not even ranking in the top 20. On a department of transportation right-of-way that in former days had been designated for a 10-lane freeway, I saw two acres of vegetables, fruit trees, and beehives. It was feeding the neighborhood and providing meaningful employment, sandwiched between an expressway and multi-tiered parking garage.

A few blocks away I visited a 1/12 acre urban farm operated by about five visionary young people. Located on a former lot where a pair of defunct townhouses had been bulldozed, these enterprising young people were feeding 20 people their year-round food requirements (except for meat and dairy) from that tiny lot. The subsoil was asphalt, wires, old apartment piping. When they dug holes to put in pea vine trellises, their holes went through blue jeans and soup cans.

These young people arranged for local chipper crews to bring in all their chips, and over the course of a few months literally built 12 inches of soil. That way the plant roots did not extend into questionably toxic subsoil. A simple hoophouse sheltered a kitchen for communal meals. A tiny quasi-legal chicken yard recycled the neighborhood's kitchen scraps. While I was visiting a couple of neighbor boys lugged over some trash bags full of kitchen scraps. That reduces landfill material. The chickens lay eggs until they get too old, then they can be eaten as stewing hens.

These young people were conducting canning and food preservation classes for folks in the neighborhood. Their earthworm beds and vibrant plants testified to ingenuity and deep ecology. In fact, other people were coming to them offering them vacant lots to expand their farm. I asked these young people how much of the St. Louis required produce could be grown within the city limits, and without hesitating, they chorused: "Every bit of it."

A few blocks away, I went to a restaurant growing most of its vegetables in a garden right outside the back door—on what was once a Kroger parking lot. The beautiful plants, black soil raised beds, and carefully constructed compost bins testified to elegance and a green thumb.

Make no mistake, if the financial district of New York were abandoned tomorrow, within a century you would scarcely know it was there. The pavement would begin to crumble and crack, then the roofs would leak, concrete would crack, and gradually everything would be taken over by trees, vines, and grass. I'm not saying I would like to see that; all I'm saying is that forever is a long time. And just because something looks the way it does today doesn't mean it will look like that a hundred years from now. Few things ever do.

The next time you begin pointing your fingers at farmers, blaming them for developing or altering that pristine pastoral landscape, ask yourself: "Would I like someone to intrude on my investment portfolio this way?" Once you settle that question, then let's see if the same restrictions you want to put on my wealth, or the same oversight you want to give yourself over my wealth, is the same you'd like on your investment portfolio. Oh, you wouldn't like that? I didn't think you would.

But you will argue: "Land is different. It's out where I can see it." I could just as easily respond that your money patronizes things I don't like. I think pornography is evil and a blight on the culture. But do you want me to lobby for regulations that will prohibit you from investing your money in pornography?

But you say: "People abuse land. That shouldn't be allowed." Now we get to even more thorny issues. What is abuse? Who defines abuse? It's kind of like the word violence. I really get tired of people using the word violence, as if it's an inherently bad thing. You want to see violence? Go to the Serengetti and watch lions interact with wildebeests.

Violence is the government sending out an armed enforcement officer to take my lifetime's accumulated wealth. Violence is inheritance taxes. Violence is Monsanto threatening open pollinated grain growers with legal action if Monsanto's GMO pollen drifts over onto their open pollinated fields and impregnates some corn. Violence is what I will do if someone comes into my house and attacks my wife and children. That is wonderful violence. In fact, I think sometimes we need a lot more violence. When Jesus took a whip to the money changers in the temple, I don't think He was smoking a peacepipe.

92

What will it take for couch potatoes to realize that the government assault against raw milk, homemade noodles, and compost-grown tomatoes is an intolerable violence? I'm with Solomon: there are times for violence and times for peace. A time to fight and a time to quit fighting. A time for mercy, and a time for punishment. Different times call for different measures. Forever is a long time.

I find it fascinating that what the preservationsts want to lock up forever due to its appearances, was actually created by a lot of landscape altering. In fact, I know a farm where a Native American burial area and village campground used to exist just a few short centuries ago. And now it's preserved as farmland. I really think it should have been preserved as a Native American community and a buffalo grazing ground. You see, when you start backing up to yesterday's evolution, it makes predicting the future absurd.

To preserve farmland without preserving farmers is like trying to keep car manufacturers without auto workers. You can't have one without the other. If you just lock up farmland without preserving farmers, you've just created a nonproductive wilderness. That threw some of you for a loop, didn't it? Nonproductive? What do you mean nonproductive? It's growing trees, isn't it?

So how many trees do we need? If we really want to sequester carbon, we need to be growing grass. And if we really want to grow grass, we need to be grazing livestock. And if we want to graze livestock, we need farmers to care for the livestock and make sure they move around to metabolize the greatest amount of solar energy into decomposable biomass. In 1820 Vermont was 80 percent open land and 20 percent trees. Today, it's 20 percent open land and 80 percent trees. Our county has 50 percent more trees than it did in 1860.

I certainly like trees. But there is nothing ecological about nursing home forests, where the dead and dying are everywhere. If you really want a hands-off, non-human interventionist policy, such areas will eventually burn in a violent (there's that word) conflagration, pumping millions of tons of CO_2 into the air. The violence, or disturbance, restarts the next productive phase: grasses, then shrubs and briars, then trees. While I deeply appreciate that the hand of man has not been beneficial to the environment in many areas, that doesn't mean it has to be so. Plenty of places on this great earth showcase the environmentally benevolent and creative hand of human intervention.

I belabor this point because in my multitudinous interactions with environmental science classes, nature campers and the like, as soon as I talk about putting people back on the land, I'm assaulted with the assumption: "But people are bad for the land." Somehow environmental teachers need to get off the negative and start teaching some positive. Why can't environmental instruction include examples of good behavior instead of just bad behavior? I believe the environmental movement in many ways is really a people hating movement, and a people hating movement cannot be a good thing.

In my opinion, the greatest resource degradation in rural America in the last century has not been erosion, pollution, or CAFOs, as bad as all of those are. I believe it's the people loss. The people who lived here, grew up here, had grandchildren here, attended church here, and spent money here. The fact that farmers don't even merit a category on the U.S. Census should create national panic. But it causes nary a whimper. In fact, as a culture we're rather proud that so few people need soil themselves being farmers.

When we begin cutting redwoods, people revolt. When we build resorts on beaches, people revolt. When we put theme parks on Civil War battlefields, people revolt. But we've lost our greatest repository of self-sufficient food production, and nobody even gives a rip. And the people who could are not interested in saving farmers; they're only interested in preserving some idyllic pastoral Normal Rockwell scene without realizing that that scene was created by a farmer.

A farmer grubbed out tree stumps to make that field. A farmer built that picturesque barn. A farmer kept that field from returning to forest. A farmer kept the blackberries and brambles from overtaking the pasture. A farmer tended those cows. A farmer planted that corn and that barley. A farmer harvested that hay that complements the barn. That landscape you want to preserve was created and sustained by a farmer. Without the farmer, it will look, in 100 years (which is a lot shorter than forever) just like Wall Street would look in 100 years without traders.

Assuming that we can sit down today, look at a piece of land, and know what it should look like even 10 years from now is both naïve and presumptuous. Let me explain it a little different way. In the 1950s and 1960s, the value relationship between farmland and raw production were completely different than they are today. In 1961, Mom and Dad bought this 550 acre farm, with the house, an equipment shed, a barn, tractor, baler, rake and hay wagon, for $49,000.

I can hear the air sucking in as you gasp at that figure. You might want to take a look at the figure again. And yes, the decimal point is in the right place. At that time, weaned calves were bringing 30 cents a pound at the sale barn. Today, according to appraisers, this farm is worth $3 million and those calves are bringing 90 cents a pound. Do you see the discrepancy here? The land has gone up 60 times but the value of its production has gone up only three times. Looked at another way, if calves had kept pace with the land, we would be getting $18 a pound. And hamburgers would cost $30 apiece.

The same inordinate ratio occurred in machinery, fuel, insurance and other things farmers buy. Insurance, net worth, and public perception are all based on a paper figure. The higher value of the land does not mean one more raindrop falls on it. That increased paper value adds not one more ray of sunshine, blade of grass, or tree leaf. At the end of the day, all that paper value inflation means is that business cannot be done tomorrow like it was done yesterday, or the farm will not economically survive. And that is the very real world of farming.

Now enter the land preservationists, who walk on the farm today, look at it today, like the way it looks today, and want to make sure it looks like this 100 years from now. Or more foolishly, forever from now. Who knows what the economic and cultural climate will be like then? The reason for this dramatic change in ratios was that in the 1950s and 1960s families cooked from scratch at home. The farmer enjoyed a large percentage of the retail food dollar. And in fact much of the processing that was done, from butchering to canning, was done either by farmers or by a farmers' extended family. It was done in the rural community and fed the rural economy.

As families moved out of the kitchen and purchased more highly processed food, the farmers' share of the retail food dollar dropped. It's still dropping. Less and less of the retail dollar ends up in the farmer's wallet, and more and more of it ends up in all the parts between field and fork: wholesale buyers, processors, distributors, marketers, and retailers. The idea that you could buy land at market prices and pay for it with the production from that land no longer applies.

That is why most purchasers of farmland in the last decade, perhaps even in the last two decades, have not been farmers. Until the Internal Revenue Service closed some of the loopholes, farming was the favorite tax write off of the 1970s and 1980s. It's still a hot one, but not quite as hot as it used to be. The e-boom created a lot of wealth that people wanted to protect from IRS tentacles. Meanwhile, suburbia was growing as urban centers went through the donut phenomenon—everybody moving out of the center. That put pressure on farmland adjacent to urban areas.

As the ability to maintain economic viability changed, farm children began leaving in droves. This left the aging farm couple on the land by themselves. And these aging farm couples were beyond the age of experimentation and creativity. The business had changed in their lifetime, but they would not change with it. Enter the environmental preservationist movement like American Farmland Trust, panicking over the loss of farmland. Interesting, this was exactly when acres and acres of land were becoming available in St. Louis. Baltimore has 40,000 acres of unused land. America has 35 million acres of lawn. That's a lot of productive capacity.

Zoning boards and local planning commissions began writing land use regulations to stop the hemorrhage of farmland loss. Saving farmland became culturally synonymous with motherhood, baseball, and apple pie. To even think about opposing farmland preservation efforts was political suicide. And this sincere effort swept the country—and in many ways is still sweeping the country—without asking the most fundamental question: "How do we save farmers?"

I know farms that are preserved so securely that they can't raise pastured poultry: the field shelters are considered "new construction." Others have dilapidated log cabin-type barns in totally inappropriate places—like a marsh. Wouldn't it be nice if the current farmer could build a new barn near his homestead on high ground? No, new construction is not allowed. Don't even think about a cannery, commercial kitchen, tourist cottage or children's day camp.

Fortunately, some groups are beginning to see the light. Recently the Piedmont Environmental Council (PEC), perhaps the premier environmentalist group in Virginia, hosted a one day seminar on leasing farmland. As the PEC organizers explained it to me, many of their members had made enough money to buy land but they were steeped in the anti-farmer mentality. Their heads filled with urban environmentalist diatribes about how farmers ruin the land, these newbie farm owners didn't want anything to do with their neighbor farmers. Those bad guys.

But after these environmentally-sensitive folks bought these farms, they realized that maintaining the beautiful pastures took a lot of mower work. Pastures had been created for, and maintained by, grazing animals. But animals are bad, so they couldn't have sheep or cows mow them. The result was that most of these new farm owners were simply mowing the acreage once or twice a year. This kept the brush and briars back, preserving the viewscape, but was a terrible way to use land. In their wisdom, the PEC staff realized that if they could link these newbie farm owners with exceptionally good farmers in their communities, the farmers could leverage their knowledge on more acreage and the landowners would not have to keep running their tractors and mowers. I hope the irony of an environmentalist running his mower over his non-farm farm hasn't escaped the casual reader.

Freezing a landscape takes a lot of work. Think about the landscape crew on a college campus. Think about what a resort goes through to maintain its appearance. How about a golf course? Keeping a landscape looking a certain way is hard. Landscapes are living things and they are changing just like your face wrinkles and six-pack abs turn into love handles. Forever is a long time.

The gathering at the PEC-sponsored landowner-farmer workshop was diverse but quite energetic. It was the first time in

America I heard the word tenure discussed as a positive thing. In Europe, tenure is common. Leases of 99 years allow multi-generational farming on land owned by someone else—historically, royalty. Some analysts believe we may be entering a time like that in America, where land ownership is for those who have already made their money and are looking for a financially defensive mechanism. Financial offense will be played by expert managers who know how to create beautiful landscapes. I know a farming outfit in China that has a 99 year lease on 1,000 acres for $50 a year. That sounds better than American ownership to me. No inheritance taxes.

Marrying farm productivity with beautiful landscapes is both doable and desirable. But that takes good farmers. And it takes ecological exercise. Not a public park or playground. When entrepreneurial savvy and sweat come together on a farm, it can be a beautiful thing if controlled by an environmental ethic.

That means the first way to preserve good farmers is to patronize them. I spoke to an exclusive male-only, membership-by-invitation-only Republican group on the west coast one morning and one of the senior members verbally took my head off when I suggested that taking the kids to McDonalds was an environmental assault. He thought I was completely over the top, and insisted that since his grandchildren liked McDonalds, it certainly didn't hurt anything to take them there. I believe McDonalds represents everything that is terrible in our food system, in our land use, and in our economy.

If all the grand matriarchs and patriarchs who fund environmental organizations would withdraw their daily patronage of industrial food, it would do more to preserve farmland than all the preservation regulations and easements put together. I'm a believer in personal responsibility. If you don't like something, clean up your own act first. But how many feel-good Sierra Club grandparents take their kiddos to McDonalds and fund Tyson chicken? Telling someone else what to do is always easier than cleaning up our own act.

So how do we preserve farmers? We patronize them. Secondly, we create a favorable tax plan. That means we eliminate inheritance taxes. Why should I have to pay the government a million dollars just to keep our farm in the family? Again, the fact that the government says it's worth X amount doesn't mean a hill of beans with regard to productive capacity. Unless I sell, it's worthless. In fact, the reason so many children are selling their parents' farms is because they don't have a clue how to make the land economically viable.

Land in and of itself does not make a viable business. It can be a hedge against economic downturn. It can be traded up or down. But owning a chunk of land does not make it a business. Land by itself does not provide for its upkeep. Saddling farmers— or anybody, for that matter—with inheritance taxes is violence against innocent bystanders. Death should not be a taxable event.

Beyond inheritance taxes, however, are real estate taxes. Here again, PEC did groundbreaking research in the 1980s showing that on average, government services cost 70 cents on the dollar in taxes generated from agricultural land. Business and industry tax-to-service ratios were about the same. But residential taxes generated only a dollar for every $1.30 in government services. In other words, business and agriculture subsidizes residential on a prorated income-expense basis.

If preservationists really want to preserve farmland, why not attack this inequitable real estate tax situation first? I think the reason is because most preservationists also like bigger government. Cutting taxes isn't on the agenda. But saddling farmers with more off-farm regulation and oversight is. I can't endorse any preservation group unless first it attacks this inequitable tax burden. The reason for the difference, of course, is that machines and cows don't need schools and jails. But neither do they vote. The result? Residential real estate has garnered for itself a very comfortable subsidy for life—paid for on the backs of farmers and businesses. If a community really wants to preserve green space, then change the tax ratios so green space is not subsidizing residential housing. This inequity is growing each year; it's not decreasing.

Third, quit subsidizing and giving sweetheart deals for development. Almost every locality has an Economic Development Authority, or Industrial Development Authority. These entities have public staffers to work out sweetheart tax, investment (like floating tax-free bonds) and infrastructure deals to attract development. When is the last time you heard Farm

Development Authority? Coming into our county on the interstates you'll see signs lauding this as a "Business Friendly" area. When is the last time you saw one touting its "Farmer Friendly" persona?

If I want to build a value added farm building like a little store or cannery or kitchen, I don't get tax free bonds and staff help down at the county building. I get vilified by preservationists, laughed at by the banker who knows farmers are a dying deal, and harassed by building inspectors and zoning administrators. Goodness, I can't even sell a neighbor's pumpkins without a permit. I'm certainly not against business and development, but it should carry its own weight just like a farmer does. Why is one big industry or development more valuble to a locality than the cumulative value of several farmers or home businesses?

Fourth, allow and encourage farmers to value-add on premises. I'll talk a lot more about this in the chapter on the village economy, but suffice it to say we need to rebuild the processing infrastructure that existed prior to centralized industrial food. Canneries, bakeries, abattoirs, wood shops. Our culture is becoming aware of localized artisanship. It's a burgeoning awareness, compared to the 1950s and 1960s when TV dinners were the new food fad.

Farms and farmers need to be appreciated for the sweat equity they bring into a community. What they look like will be different in a localized economy than what they looked like in a pre-industrial economy. The land needs to be used, to be reformed and transformed to be more productive, to cycle water better, to convert solar energy into biomass better. That will require people, houses, biodiesel and wind energy facilities. That will change the landscape. But it will be better. The wildlife can and should still be part of it. Our farm has far more wildlife now than it did 50 years ago, before ponds, thick grass, and roads.

Exercising the land is part of the sheer ecstasy of being a lunatic farmer.

TAKEWAY POINTS

1. Landscape exercise tones it to greater health.

2. Forever is a long time.

3. Preserving land without preserving farmers is foolish.

4. Tax prejudices against farmers threaten their viability.

Chapter 9

Normal Food

At our farm, we produce normal food. That means historically normal. Normal like centuries ago. That means it grows like, looks like, tastes like, and handles like the same stuff our great grandparents grew and ate.

Memories are unbelievably short. When pundits decry political rhetoric and accuse one side or the other of encouraging disunity, I wonder what they would have said about duels and canings that politicians engaged in during earlier times of our country's history. That's why the study of history is important. It gives us perspective. In reality, we as humans just seem to think that whatever is must be normal.

I think this memory myopia is especially acute in our day of techno-glitzy innovation. The speed of change is giving us an inordinate sense that something just a decade old classifies as old-timey. Cars can now be registered as antiques when they hit just 30 years of age. Would anyone think a person 30 years old is an antique?

This dissonance between technological and biological change is important because in our technology-worshipping modern world we assume this level of accelerated change, this warp speed newness, is normal. Statisticians say we're living in the hockey stick age. Every trend line, from population to medical costs to energy consumption looks like a hockey stick. Hockey stick trend lines do not last long.

As Gordon Hazard, iconic Mississippi grass farmer, says: "I haven't seen anybody throw a ball so high it wouldn't come down." But it's easy to get caught in the euphoria of new. We can easily get drunk on today.

I am not opposed to change. I'm certainly enjoying writing this book on a computer much more than I enjoyed writing my first book, *PASTURED POULTRY PROFITS,* on a typewriter. I certainly enjoy building compost piles with a front end loader more than shoveling by hand. But to give in to intoxicating desires for more and bigger and faster is to assume that hockey stick projections are normal.

And especially when it comes to biological systems, change is much slower than in mechanics or technology. Tomatoes have not changed much for a long time. Potatoes still grow in the ground. Cows still have four legs. Compared to travel, for example, in a century that took us from horse and buggy to space capsules, technology is in an acceleration lane all by itself.

Lest you think today's corn is as different from corn a century ago as the horse and buggy is from space capsules, it's not. If you could bring Thomas Jefferson back to his beloved Monticello and show him today's corn, he would have no trouble identifying it as corn. Show him a space capsule, however, and he would probably have a hard time identifying what it was. That's my point. Biological changes are in a whole different league from technological.

That means that our internal digestive community of 3 trillion organisms is on its own time frame. These critters that inhabit our intestines live independently of space travel and computers. The critters that live in our intestines today are just like the ones who lived in Thomas Jefferson's. And they want the same things.

They don't know about the religious right or the liberal left. They have no clue about Republicans and Democrats. And they certainly don't know about Twinkies, Cocoa Puffs and Mountain Dew. These are totally foreign substances. These bacteria don't know how to communicate with high fructose corn syrup. A car designed to run on gasoline would have a rough time if you poured diesel fuel into the tank. If a simple man-made machine is that picky, think about the community of critters inhabiting our insides and the intricate balance, relationships, assimilation and excretion that must occur there. It's mind boggling. And the more you study it, the more mysterious it becomes.

We farmers are as guilty as anybody of defining normal by today's status quo. Often I want to go Times Square, get up on a soap box (are you surprised?) and yell to the world: "Folks, this ain't normal!"

It's not normal to apply super triple phosphate to plants. It's not normal to apply anhydrous ammonia to the soil. It's not normal to eat food that you can't pronounce. It's not normal to eat food that you can't make in your kitchen. If you went to the average supermarket and removed everything that would not have been there in 1900, everything except the outside shelves would be empty. The outside shelves contain the produce, meat, dairy, and bread. The inside aisles contain soy and corn syrup plus something else. Have you ever tried making corn syrup in your kitchen? You need a laboratory to do it. How about Red Dye 29?

I've heard that we shouldn't eat any food that wasn't available before 1900. If we only applied that test to what we eat, it would completely change the way Americans eat. I'm so thankful that hot dogs were introduced at the 1890 World's Fair.

Many folks today can't imagine family style sit-down home-cooked from scratch meals. A friend recently told me she was retiring as a high school art teacher. I asked her why, and here is what she said: "Every year I ask my 10th graders to bring in a cooking pot as a beginning drawing exercise. This year, out of my 20 students, not a single one had a cooking pot at home. I asked them: 'What do you cook in?' And they said: 'We just put the box in the microwave.'"

Folks, that's scary. You know something? That ain't normal. But we have an entire food system, from production to processing, that thinks our civilization is normal. Or at least enlightened. So much so, in fact, that other cultures should emulate us. We think our normalcy should be exported around the world so all those people and cultures can become like us.

104

As a culture, we are running a giant experiment. Individually, we are running a giant experiment on ourselves. How much of this abnormal fare can my body stand before it revolts? Make no mistake about it, the escalation of Type II diabetes and obesity are directly linked to this giant experiment. The longer we let this experiment run, the more red flags it will wave. Recently I saw that Congress was contemplating a tax on beverages containing high fructose corn syrup to curb its consumption. Instead, why doesn't Congress quit subsidizing corn? If they really want the price to go up so people won't drink as much, stopping subsidies would have the same effect and the money saved could be spent on whatever they were going to spend the tax on. Or how about this?—lower taxes instead. Now there's a novel idea. Letting workers keep some of their own money. Novel indeed.

The whole clean food movement, amazingly, has defined itself as unconventional. As if it's conventional to spray pesticides. As if it's conventional to knife anhydrous ammonia into the soil. As if it's normal to sterilize strawberry fields with fumigants. As if it's normal to confine 15,000 chickens in one house. As if it's normal for the average morsel of food to travel 1,500 miles from field to plate. Goodness, the average T-bone steak has seen more of America than the farmer that raised the steer.

I was in a hotel the other day exploring the complimentary breakfast. Nothing was edible. Come on: donuts, margarine, pasteurized milk, Pop Tarts. I spied a bowl of icy water with Yoplait yogurt cups floating. Ah, some food. I looked on the label and one of the first ingredients was high fructose corn syrup. Why do they have to put high fructose corn syrup in yogurt, for crying out loud? This adulteration of the food supply is both unconscionable and unprecedented.

If we could speak to the bugs in our bellies, and ask them what they'd like to eat, I don't think they'd respond: "Whatever Monsanto concocts is fine with us." I think they would say: "What we've been eating since creation. Geographically and culturally diverse, yes. But dissected, genetically prostituted, chemicalized, irradiated and reconstituted, no."

Perhaps it would be a good exercise just to brainstorm some of the abnormal things in our food system in order to appreciate the magnitude of the departure from historical belly bug normalcy. This is not exhaustive, but it does offer springboards to discussion. All of this has happened, for the most part, since 1900. The format here will be a listing of the current abnormality, then the historical counterpart, and in many cases new lunatic alternatives.

1. Acidulated chemical fertilizers. From slash and burn, to fallow, to the 7-year rotation, indigenous systems have relied on rotations, recycling, and relationships (plant to plant, plant to animal, animal to animal) for fertility management. Minerals came from seaweed or ground rock powders. Things applied to the soil would not burn your skin or kill earthworms. From about 1920 to 1940 Sir Albert Howard introduced the world to the science of aerobic composting. If all the excrement, plant residue, and offal waste stream were composted, chemical fertilizers would be totally unnecessary. Today, in addition to efficient and scientific composting systems, we have foliar feeding and manure tea as a fertility supplement.

2. Organophosphate pesticides. Interestingly, if you study pest losses as a percentage of crop losses before and after pesticide use, it is not better today than it was a century ago. In aggregate, crop losses as a percentage of yield have remained constant. Today's high tech biologicals and knowledge of good rotations and degree-day monitoring aid planting strategies to minimize risks.

3. Grubicides and parasiticides. Diversified farms offered lots of checks and balances. The reason for the proliferation of diseases in the first half of the 1900s was that farmers were amalgamating animals in large groups ahead of pharmaceutical advances. The proverbial slinky effect. It took awhile for pharmaceuticals to catch up with the CAFOs. The bugs are cleverly outrunning the pharmaceutical development today, but we have Eggmobiles, intensive rotational grazing, pastured poultry, electric fencing and a host of techno-glitzy stuff to raise commercial-sized groups away from buildings and muddy barnyards.

4. Genetically modified organisms. Heritage breeds developed due to careful observation and selection, over time.

Gradually a phenotype emerged that was productive and adaptive to that bioregion. Today we have line breeding and continued genetic selection to create nativized genetics for localized adaptation.

5. Irradiation. Zapping food to sterilize it alters its cellular structure. Heat used to be the cleanser of choice—boiling water, autoclaves, cooking, canning. And carcasses were processed one at time instead of in heaps. How satisfying to know that at least when we eat poop now, it's sterile poop. Today, of course we have stainless steel, rural electrification, temp strips and bacteria monitoring capabilities. Not to mention UV lights to purify water.

6. Confinement Animal Feeding Operations. Animals used to be spread out, exercising, breathing fresh air, getting sunshine. A lot of people used to grow things—even if they lived in town. Today, with electric fence and symbiotic production systems, CAFOs are completely obsolete.

7. Long distance food transportation. Certainly tea, spices, and coffee transport have been around for a long time, but staples have always been grown close to home. Home preservation, season extending cold frames, and root cellars were part of the homestead. Today we have farmers' markets, Community Supported Agriculture, local distribution companies, virtually free electronic communication that stimulates networking, freezers, solar food dryers.

8. High fructose corn syrup. Historic sweeteners were cane sugar, molasses, honey, and maple sugar. They still work today, and if more people patronized these historic sweeteners, the infrastructure and production would grow to meet the market.

9. Non-water drinks. Who needs them? High tech purification systems now offer nature's hydration system—good old plain water. Most people should drink at least three times as much as they do.

10. Eating-on-the-run. Remember the family sit-down evening dinner? Some 50 percent of all meals are now prepared outside the home. Almost one in four meals is eaten in an automobile. The mealtime repast has been replaced by nibbling, snacking, and boxed grazing.

11. Highly processed foods. Back in the day, domestic culinary arts dominated in all kitchens. People knew how to cut up

a chicken, make applesauce, and bake bread. Today's resurgence of food shows and culinary skills marks a delightful comeback for these arts. This time, though, dicing and cooking gadgets up the wazoo have revolutionized heritage procedures. Slow cookers and crockpots have replaced the fireside vigil.

Whenever someone asks me: "What would be your one piece of advice for my community to reduce industrial food?" my answer is always the same: "Quit buying it and begin cooking from scratch at home and dining together. And turn off the TV while you're eating, so you can learn again to converse around the table." That advice is as appropriate in our hamlet of Swoope as it is in the most upscale block of New York City.

Nobody is putting a gun to anybody's head and making them buy Cocoa Puffs or frozen pizza. The opt out alternative is real and still the most powerful way to disempower things we don't like. Just take away their funding. Stop patronage. Vote with your pocketbook. If we plagiarized the Great American Smokeout and did a Great American Fast Foodout, it would bring the entire industrial food system to its knees.

Every day, I'm told, five tractor trailer loads of French fries enter Washington, D.C. Can you imagine? What would happen if for one day nobody ate at an industrial food-supplied burger joint? Talk about shaking up the system. For all its apparent strength, the margins in the industrial food system are literally fractions of pennies. The volume makes up for the tiny gross margins.

Just one day, and certainly two, of boycotting fast food would discombobulate the entire system. Wouldn't it be fun to have the entire industrial food system on its knees begging for forgiveness? Asking: "What must I do to be appreciated?" Wouldn't that be fun? Talk about ecstasy. And the beauty of it is that this action does not require legislation, grants, government agencies, organizations, or more taxes. It would be a hoot if the industrial food system got whacked in the pocketbook with the notion that ultimately the customers are in charge. That would be a revelation. Something industrial food elitists probably never thought of before.

Whenever anybody starts whining that some sectors of the population don't have a choice, I ask them: "Do any of these people have a TV? Do any of them drink soda? Do any of them have iPods and eat Twinkies for breakfast?" I could go on in this vein, but you get the drift. Several years ago the local Food Bank called us to see if we would take their throw-away food and feed it to our pigs. We agreed to try it.

We went weekly and brought home everything from lettuce to five gallon buckets of broken eggs. One week we brought home two tons of premium beautiful sweet potatoes. Now folks, I love sweet potatoes. This haul filled up two heavy duty pickup trucks, all the way to the top of the cattle racks. Of course, the pigs loved them, but we felt immoral for feeding them to the pigs. We ate all we could for breakfast, lunch, and supper.

The following week when I went to pick up stuff, I asked the staff about all those wonderful sweet potatoes. Without batting an eye, they said: "Oh, that's poor man's food. Sweet potatoes are poor man's food. These people want Ho-Ho cakes, sodas and canned stuff. They don't want to fix anything." At the risk of sounding callous and hard nosed, these folks don't have a hunger problem. They have an initiative problem. And that isn't helped by just pouring more processed food down their gullet for free. Whew! That will get me crucified in some quarters, I'm sure, but somebody has to say it.

Nobody is better at whining than college students. A local college professor teaching a global nutrition class asked me to come and lecture to her students. The professor and I colluded to create a powerful lesson. I brought eggs, local cold-pressed apple juice with an inch of sediment in the jug, and skillets. The professor found local salsa, some wonderful cheese, and butter.

When I showed up at the class, we fired up a stove and I cooked omelettes. The students chose their omolette stuffings and within a few minutes they were cooking and going after it. It was delightful. My omelette cooking is based on a chef representing the National Egg Board that I enjoyed watching at a North American Farmers Direct Marketing Association annual meeting in Phoenix many years ago. He did a 60-second omelette. The key was putting about a tablespoon of water per half dozen eggs to percolate up through the beaten eggs helping them cook easily and stay moist.

Instead of flipping it over in the pan, he essentially cooked it like a glorified scrambled egg, but put in the additives on half of it while it was still in the pan. Then he flipped the other half over the top and served it by turning it over on your plate, upside down. That way even though the bottom might be torn in the pan, the top was always perfect because it was the unmolested underside. Works every time. A big hunk of butter melted in the pan before each omelette keeps things from sticking.

Anyway, when it was all over, the meals took about 120 seconds (2 minutes) to prepare and cost $1.50. We had dined like kings. I asked the students: "How many of you have watched a movie this week?" Every hand went up. There's 110 minutes on average—enough time to fix 55 of these meals. Not willing to let up the pressure, I continued: "How many of you have drunk a soda this week?" Again, every hand went up.

That's 75 cents at least, and many times more. I pressed on: "How many of you have put anything in the vending machine this week?" Oh, my goodness. Lots. Crackers. Munchies. By this time, they all had a sheepish look as the collective epiphany sunk in: we make time and find money for what we really want to do. It's a lot easier to whine and say "I can't" than it is to get out of our routine and make change.

All new behavior takes time to learn. Domestic culinary skills have not been honored for quite some time. Liberating women from the kitchen, from home and hearth, was supposed to be a good thing. But now as that pendulum has swung too far, many of us realize that a dark underside exists from this anti-kitchen mentality. Just like we need to resurrect an honor and respect for our farmers, we need to honor and respect domestic culinary skills.

I've been greatly encouraged the last couple of years when I speak on college campuses. Before I do the classroom lecture, the students bring in potluck food. It's wonderful. Sometimes it's food they've grown in a community garden or campus garden. Sometimes it's food they acquired at the farmers' market or through a CSA. In any case, it's a revival of interest in local sourcing, in-home cooking, and fellowshipping around a meal. Now that's sheer ecstasy. And certainly considered lunacy by the industrial food crowd that wants people to be disempowered and victimized.

On that note, let me share something that's been helpful for me when confronted with new thinking or the challenge of new skills. We've all heard the old saying: "If it's worth doing, it's worth doing right." Let me tell you something—that's wrong. The way the saying should go is like this: "If it's worth doing, it's worthy doing poorly first."

Many of us, when confronted with something new, suffer a paralysis of fear. New things are often terrifying. But all new things take time to learn. What parent says to a toddling child: "Johnny, if you can't walk any better than that, just quit. That's ridiculous, falling all over and crying and bumping your head against the wall. If you can't walk any better than that, quit trying." No, instead the parent encourages the child, cheers him on. "That's great, Johnny, you're getting the hang of it. Stay with it. Look, everybody, Johnny is starting to walk."

As we get older, however, new things carry a stigma. "What if I mess up? What if I can't get the hang of it? What if it takes too long to learn? What if I have to give up something else in order to do this? What if in the end I don't like this?" A host of what ifs will cloud our heads to the point that we never attempt something new. To many people, what I'm suggesting—find your local food and cook it at home—is as scary as window washing a skyscraper. But you don't start at the 100th floor. You start on the ground.

You get the hang of it with simple things. You make a hamburger. You slice a tomato. You fry an egg. The other day I was in the supermarket and saw precooked microwavable heat-and-eat bacon in a box. Have we really gotten so busy that we can't even fry bacon?

I spoke at the University of Guelph many years ago and one of the other speakers was a high powered attorney from Toronto. She lived in a sixth-story condominium, knocked down a six-figure salary. Then she and her husband had a baby. All of her maternal instincts kicked in and she decided she wanted her baby healthy, which required eating the best food possible. Initially, she decided to breast feed, which of course her peers thought was lunacy. But the new mom decided breast feeding was one of the ultimate personal empowerments, a great opt-out strategy to the industrial food cartel.

I have thought about that decision a lot. Probably the decision, in the American culture during the 1950s and 1960s, that breastfeeding was Neanderthal and barbaric, was the low point in abnormality. In the history of humans, this was unprecedented. So we raised a generation of asthmatics on Infamil and Similac. What a terrible resource waste. All those breasts out there not being used, and instead all that value going to artificials. Tragic indeed.

Anyway, this attorney and her husband, as the next step in their opt out strategy, decided to take all their entertainment budget, both time and money, and spend the next year searching out their food resources. They would forego their vacations, Blockbuster videos, whatever. They would go on a treasure hunt around Toronto, sourcing their food from farmers with the goal of ending the year with no bar codes in their pantry. And to the amazement of those Guelph students, she exulted: "We did it!"

I'm not suggesting that if you have barcodes in your pantry you're in sin. But here is a new mom in a big city with hurried harried lifestyle who decided to not be a victim. Let's applaud her, put her on a pedestal, emulate her. If everyone followed her lead, we wouldn't even have CAFOs and irradiation. She took the power of one to its ultimate meaning. And prevailed. That's why I tell folks who whine and complain because they think they are trapped by the system: "Go on a treasure hunt. Every single area is rimmed with farm and food treasures. Take your deer rifle and TV up on a hill, put a bullet through it, and invest all that time in discovering your nearby food treasures and your own kitchen."

Saying these kinds of things is certainly not politically correct, but it sure beats sitting down with the whiner and having a pity party together. Listen to some motivational speeches. In life as well as in business, the race does not go to the whiners. It goes to the folks who run harder and sleep less. Whose fire in the belly translates to practical life changes.

Food police aren't normal. People have always been able to eat pretty much whatever they wanted. No civilization has ever had bureaucrats determine for the populace what is and is not acceptable to eat. As the industrial backlash against local and normal food escalates, it will be interesting to see how much good food gets demonized before normal food wins the day. I have no doubt that normal food will again be legal; the question is how long and how far-reaching will be the assault until our side wins the day.

The point of the assault is raw milk specifically, and raw dairy generally. By what strange process has raw milk been deemed unsafe but Twinkies, Cocoa-Puffs and Mountain Dew been deemed safe? This convoluted circumstance derives from the notion that unless something kills you outright, it's safe. If it gives you Type II diabetes or asthma or some other chronic malady, that's okay because it didn't happen quickly enough to go on the causation report. Scientists use the lethal dose rule to measure toxicity. It's a crazy notion. The only things considered harmful are things that kill you while the study is going on.

I won't belabor the issue in this book since I wrote *EVERYTHING I WANT TO DO IS ILLEGAL* to address it in depth. But here are a couple of new thoughts since I wrote that book. When the government gets between my lips and my throat, that's what I call an invasion of privacy. Maybe you don't see it that way, but I do. I am amazed that people who squawk about parental notification in order for underage women to get an abortion don't get upset about being denied raw milk or yogurt from a neighbor dairy farmer. Surely getting raw milk is an easier freedom of choice issue than abortions.

Fortunately, our side now has a wonderful legal advocacy group modeled after the Home School Defense Association, which paved the way for legalized homeschooling nationwide back in the early 1980s. The Farm to Consumer Legal Defense Fund (FTCLDF) was founded in 2007 by visionaries working with the Weston A. Price Foundation (WAPF). Founded in 1999 by Sally Fallon, WAPF now has memberships and local chapters all over the world dedicated to preserving wise traditional diets.

The FTLCDF offers subsidized real time legal counsel to farmers being harassed by food police. Several high profile cases are highlighted in David Gumpert's 2009 release: *RAW MILK REVOLUTION.* If I may step out here where angels fear to tread, I contend that the domestic terrorism the U.S. government is waging against its own people, whether they be alternative health care professionals or backyard farmers, is certainly just as big a story as any terrorism occurring in foreign countries.

Take the case against Mark Nolt, old order Amishman in Pennsylvania, who refuses to take a license for selling his raw dairy products to his customers, citing the 14th amendment, the right of private contract, as his legal authority. Government agents have raided his dairy several times, taking product, destroying equipment, and in one raid stealing my book *EVERYTHING I WANT TO DO IS ILLEGAL* from the on-farm store's bookshelf.

In the Bible, Romans 13 describes very plainly the role of government: it is to be a "terror...to evil" and a "minister of God...for good." That is as succinct as it gets. Governments are supposed to be a terror to evil and an encourager of righteousness. But here is Mark Nolt, a pacifist, non-voter who doesn't even have a Social Security number, milking his cows and offering goods to people who voluntarily opt out of government-sanctioned fare, being handcuffed and marched off to the courthouse in police cruisers. Government officials, guns at the ready, dismantle equipment and destroy property.

This is symptomatic of more and more sting operations in which any rational, freedom-loving person comes away realizing that our government has become a terror to righteousness, and an encourager of evil. That is a sad state of affairs. I confess that it's hard for me to appreciate what our military is doing abroad when our own government agents are terrorizing our populace at home. In my opinion, the real heroes of our day are the everyday farmers who feed their neighbors with normal, historically-accurate food and take it in the teeth from industrial naysayers, food police, and consumer lethargy.

Anyone can sign up for a free education, meals provided, the best weaponry and communications money can buy to go empire-build on the other side of the world. It takes a real visionary, perhaps even a lunatic, to embark at home, where he grew up, among neighbors, on a journey that the culture thinks is lunacy. To stand with nothing but the power of righteousness. To continue massaging the land, massaging the animals, massaging the plants, nurturing in humility and awe, when the culture applauds dominion, manipulation, disrespect, and rape. I say, give me more Mark Nolts. I'll take my stand with him any day. It's a stand of principle. He is the true patriot.

Our Virginia Governor Timothy Kaine, just a few months before completing his term of office, came by our farm for a tour. Totally engaged and delightful, Kaine eventually asked me how I dealt with the industrial food system. After the whirlwind farm tour, he realized just how different this farm was from the ones he normally visited. He wondered how our farm held up, how we interacted, in such a hostile farming environment.

Here is my answer: "Governor Kaine, Monsanto doesn't scare me at all. And even though the Virginia Poultry Federation considers me a bio-terrorist because my chickens commingle with Red-Winged Blackbirds who supposedly take my diseases to the science-based environmentally-controlled Tyson chicken houses and threaten the world's food supply, they don't scare me. But they do dislike, even hate me. If what I do became normal once again, it would completely invert the power, prestige, and profits of the current food economy, from producer to retailer.

"But they can't come after me with guns. All they can do is argue with me. And they know it. So they will use you, sir, to move their agenda forward. They will stroke you with wine and cheese dinners. They will give you money. They will give you honor and plaques and recognition. They will use you to terrorize me. Because you, sir, can send people with guns.

"At the end of the day, governor, you are the thin veil of protection between the industrial food agenda and me. You stand between me and annihilation. You and every other elected official must understand that your responsibility, your ministry, your number one job, is to defend righteousness against an evil agenda."

I implore, I beg, I plead with any elected official reading or hearing these words to understand the gravity of the situation. Never, ever, in human history have we had a food police to deny the populace normal food. Never. To swoop down on little cottage food industries with legal paperwork and litigation threats just because a letter on a label is bigger or smaller than some bureaucrat thinks it should be—this is not normal. It's never been tried.

When will people begin to realize that the unprecedented acceleration in obesity, Type II diabetes, and food-borne pathogen illness is a direct result of the food police denying the populace normal food?

This whole issue of what is normal creates a conundrum for a communicator like me. I hate it when I say we're in the alternative farming business. Or we have alternative food. That's wrong. The Tyson chicken house is an alternative. The fumigated strawberry farm is the alternative.

As a way to preserve my own communication sanity, I've resorted to calling what we do normal. That forces the other side to defend itself on the basis of a historical record. I don't know about you, but if I were a gambling man, I'd bet that our 3 trillion bugs in our bellies have been eating raw milk a lot longer than they've eaten high fructose corn syrup Yoplait yogurt. I'm putting my money on the horse that's been around for millennia, instead of the one that's only been around for a few decades.

Rediscovering normal should be a dream for everyone. And I'm not so naïve that I diss all mechanical, technical, medical and hygienic discoveries. Not at all. But when it comes to foundational biology, if our internal community is not getting something that looks like, tastes like, feels like, and nourishes like what it's used to, we'd better be pretty sure about the discovery. The fact is, we're not. And we're already seeing the signs of our hubris. It's time to approach our belly bugs humbly. And beg forgiveness.

You see, I'm the one that's normal, after all. And that's the sheer ecstasy of being a lunatic farmer.

TAKEAWAY POINTS

1. Eating food you can't pronounce is not normal.

2. If it's worth doing, it's worth doing poorly first.

3. Whiners and victims don't solve problems.

4. The role of government is to encourage righteousness and discourage evil.

Respect for Life

Chapter 10

Pigness of Pigs

P erhaps no sound bite I've developed captures the essence of my sheer ecstasy and lunacy as when I say here at Polyface we want to preserve the pigness of pigs. By extrapolation, of course, this includes the cowness of cows, the tomato-ness of tomatoes. What it means is respecting and honoring the essence of the being and creating a habitat that enables full physiological distinctive expression.

Wow. That's a mouthful. In my view, this idea has its roots in the Bible. The creation record in Genesis starts with the imperative for each biological life form to reproduce after its kind. Those are two things: reproduction, and uniqueness (after its kind). Actually, the after its kind imperative means genetic similarity to the parents. In other words, you don't have pigs making baby calves and tomatoes making baby oak trees.

Forgive my simplicity, but for me, this by itself is enough to question genetically modified organisms. Many if not most GMOs do not even reproduce because they are sterilized in order to insure that buyers purchase from the seed company every year. This is to preclude seed saving. If anything speaks to food security, seed saving certainly does. When people can't save seeds, they have become slaves to outsiders.

Secondly, "after its kind" speaks to genetic order and distinctiveness. Blurring genetic uniqueness creates brand new life forms that violate the clear Genesis mandate. We now have tomatoes that are part potato, part pig, and part human. This whole genetic confusion creates big problems. When a pork gene is inserted in a tomato, what does a devout Jew or Muslim do? As we tear down the genetic fences between plants and animals, what happens to vegans when they eat animal genes inserted into plants?

John Ikerd, professor and preacher for sustainable agriculture, uses the cell wall as a metaphor to describe protective walls in general. Cell walls protect the cell from invasion and hold all the good parts together. Walls are not bad things. They form lines of demarcation that are both offensively and defensively protective. Yes, just like anything they can be abused, like the Berlin Wall. But a culture that assumes no walls should exist is a culture ready to collapse. Genetic walls protect DNA and purity.

You don't have to read far into the literature to realize that allergists are prophesying unprecedented allergy problems with genetic modification. A person who was allergic to tomatoes could eat peppers. And a person allergic to peppers could eat tomatoes. But now, with GMOs, people allergic to either of these cannot eat the other because tomatoes have pepper DNA and peppers have tomato DNA. That's one reason incidents of food allergies are exploding.

And for those who think genetic modification is just a techno-extension of traditional cross breeding, tell me by what hybridization program you would have tomatoes contain Angus cow DNA? I've watched a lot of critters breeding in my lifetime, and somehow I just can't get my head wrapped around that one. No pun intended. Talk about a convoluted mess.

But what about progress and scientific discovery? To question GMOs is like questioning the sewing machine and cotton gin, right? I may be wrong, but I come back to a difference between animate and inanimate. Living things are not just machines. The western world is an extension of Greco-Roman reductionist systematized fragmented individualized linear parts-oriented disconnected it's-all-about-me thinking. To be sure, this has led to huge breakthroughs.

But an equally valid world-view comes from the east: holism, connectedness, wholes, community-based, we're all relatives, and it's-all-about-us. This gives us spiritual and mental anchors, morality and ethics.

The east asks the questions and creates the parameters that keep the west from innovating things that we cannot morally, spiritually, physically or mentally metabolize. Without constraints, we clever humans can invent things we can't metabolize. Dad used to call this "overrunning our headlights." And so we invented nuclear power before we figured out how to handle the waste. We invented GMOs before we figured out how to keep the pollen at home. We invented DDT before we figured out how to keep it out of streams and rivers.

The iconic movie Jurassic Park offers great commentary on this problem. The euphoric scientist exults in his ability to bring dinosaurs back to life, supposedly within the protected and secure confines of a heavily fenced area. Even when the dinosaurs begin eating cars and people, tearing down the fences, and essentially destroying civilization as we know it, the scientist is drunk with achievement pride. The journalist gets in his face, if you'll recall, and asks the question that shapes the movie: "But just because we can, should we?"

It's a powerful question, and one that behooves us all to ask. Just because we can overnight air freight organic certified refrigerated flowers from Peru to San Francisco boutiques, should we? Just because we can press 15,000 chickens into a house and keep them alive with slow-release antibiotics, should we? Just because we can stretch methane balloons over confinement dairy manure lagoons, should we?

But in our American culture, we don't ask that question. The only question we ask is how to grow things faster, fatter, bigger, cheaper. Intuitively, we understand that is not a noble or sacred goal because if it were, we'd all aspire to be the fattest person in the room. Goodness, the reason the average NFL player dies before 60 is that when your neck is bigger than your head, you're a freak; and nature weeds you out.

To my knowledge, the United States Department of Aggravation...er, Agriculture (USDuh) has never asked, in any of its research: "How do we make happier pigs?" The American paradigm only views pigs as inanimate piles of protoplasmic structure to be manipulated however cleverly hubris can imagine to manipulate them. I suggest that a culture which views its life forms in that disrespectful, arrogant, manipulative fashion will view its citizens the same way, and other cultures the same way.

How we respect and honor the least of these creates the moral framework that supports how we respect and honor the greatest of these: people. How can we possibly expect our children not to shoot each other when our culture's best and brightest minds are employed in finding the stress gene in porcine DNA so it can be extracted, enabling us to mistreat the pigs even more egregiously but at least they won't care? Since more is caught than taught, what does this activity and scientific investment say to our young people regarding their dreams and ambitions, finding their individual gifts and talents, and pursuing their Tomness, Janeness, or Maryness?

When Americans, especially those involved in the military industrial complex, shake their heads in bewilderment over why the rest of the world does not endorse American values, this is why. As a culture, we've squandered moral and ethical values and prostituted the most distinctive building blocks of life to the highest bidder. Unless and until we curb this frenzied orgy that uses and abuses with insatiable amoral capitalist appetite, the world will continue to view us as a disrespectful and egocentric monstrosity.

I actually believe there is more to life than conquering and acquisition. How about nurturing and discovering how to live better within creation's order and plan? Why must everything be manipulated to short-term human gratification? Why can't humans learn how to live within the confines of nature's order? Why can't the path to discovering the essence of pig be as enjoyable as turning the pig into something that's not very piggy?

While I do not worship pigs, I think reverencing the essence of pig is still a noble goal. I don't elevate the pig to personhood, like animal worshippers do. I don't think a pig is a dog is a cat is a person is a fly. And for people who think eating animals is barbaric, I respond with a friend's admonition: "Everything is eating and being eaten. Everywhere. All the time. If you don't believe it, go lie naked in your flowerbed for a few days and see what gets eaten."

I have no problem with vegans or vegetarians. But when you tell me I can't have any compassion because I eat animals, we have a big problem. Even Jesus ate plenty of animals. If He had had a problem with it, I'm sure He would have brought it to our attention. This prejudicial view toward human omnivores, rather than demonstrating a new heightened awareness of cosmic spirituality and evolution, actually demonstrates an unprecedented disconnection with our ecological umbilical. The question: "How can you possibly eat animals?" actually shows spiritual devolution. The cycle is life, death, decomposition, new life. You can't have life without death.

And you can't elevate humanity if animals and people are the same thing. Unfortunately, the only relationship many people have with the living world is the one they enjoy with a pet. This myopic existence jaundices a balanced life view. And then when you realize that America spends more on veterinary care for pets than the entire continent of Africa spends on medical care for humans, the imbalance is painfully apparent. This isn't normal.

The Bible, which has probably been abused more than anything else to justify the kind of manipulative, disrespectful attitude toward animals and plants than any other historical text, actually has a lot to say about respect and appreciation towards the animals. Especially domestic livestock. "Don't muzzle the ox that treads the corn" certainly elevates animals to a position of laborer worthy of recompense. The Lily of the field is more beautiful than Solomon in all his glory. God knows when a sparrow falls. These all indicate something more than just inanimate objects worth no more respect than a hunk of clay or piece of plastic.

I am not a great cerebal academic. I'm just a lunatic farmer out here enjoying the plants and animals, communing with the most beautiful little niche in God's creation. I think it's important to share what happened when I took this pigness of pig concept to Yale and spoke in one of their graduate studies seminars. I gave a prepared monologue and then the idea was to engage in discussion for the next hour or so. Are you ready for the first question? Here it is: "So if a woman doesn't use her uterus, does that mean she's not expressing her woman-ness?" Let me ask you—were you thinking of that as you read through this section? Neither was I. Never entered my mind. That's why we have places like Yale, to pursue the great questions of humankind.

Notice I never said that a pig had to have babies to express its pigness. Forgive me for not tying up every conceivable loose end on my metaphor. The question was so far out of my thinking that I just laughed. I can't remember what I said. But these academics pursued this line of questioning to the exclusion of almost everything else I'd said. And I had discussed local food systems and compost-grown fertility and grass farming and all sorts of neat things. But they were stuck on the uterus. I'm sure glad my mind is too simple to work like this.

I thought I'd bring up this incident to show that I don't have it all figured out. And also to confess that ramifications of this idea exist that I've never thought about. When the pigs are gamboling around the pasture or enjoying their pigearating, I don't meditate about women's uteruses. I just enjoy the pigs expressing their pigness and that's good enough for me. Simple people like me seem to understand what I'm trying to say, and I appreciate it.

There is a radio show in the Midwest, hosted by two guys who love industrial food. I assume they like industrial food, because they like industrial agriculture. I find it pretty disingenuous when Rush Limbaugh applauds Dennis Avery, author of *SAVING THE PLANET WITH PESTICIDES AND PLASTIC* but then goes out to eat at restaurants that buy local and pasture-based. Avery actually lives in Swoope, Virginia—the ultimate cosmic balance. When he was telling folks free range poultry like ours threatened the planet's food supply, his own poultry was free ranging on his pond. Pretty hypocritical, I'd say. But he must not be a lunatic because he guests on major media shows all the time. In his world, multi-national corporations thrive and little businesses die.

Anyway, these radio guys asked me about the chickenness of the chicken. I told them it meant the ability to scratch, exercise, breathe fresh air, eat a diversified diet that included some bugs and grass, to be housed on new ground routinely, and to not have their beaks cut off. Doesn't that sound, intuitively, like a good thing? Not to these guys. Not to the industrial food advocates.

The reason this is important is because for years I entertained this fantasy about being on Oprah Winfrey with Don Tyson and asking him this question: "What is the chickenness of the chicken?" And in my arrogant little mind, I figured he would be stumped. As if he wouldn't have an answer. How stupid of me. These guys aren't dumb—they are sharp as can be. They didn't get to their positions by being tongue-tied when thrown curve ball questions.

So here were a couple of radio show hosts, whose audience was primarily industrial farmers, and who would provide a wonderful trial run for my killer question. Here was their rendition of the chickenness of the chicken: all the feed they want all the time, environmentally controlled housing, safety from predators, and safety from other chickens' pecking, commonly accomplished by debeaking (cutting half the beak off to blunt it).

Clearly, these guys were trying to put me in the "he doesn't care about the chickens' welfare and safety" box. Because, let's face it, when the chickens are out on pasture, one does occasionally get picked off by a hawk. And every day is not just blue skies and 70 degrees. Sometimes it rains and gets a little muddy. And with a full beak, the birds can injure each other fighting. That's all true. But notice how they defined the chickenness—it's all about protecting, being safe, and being taken care of.

Stay with me. I'm going somewhere with this. If I asked you what allows you to fully express your Maryness, for example, what would say? I'll bet the first thing out of your mouth would not be "protection from the schoolyard bully." You can express your Maryness quite well in the way you respond to the schoolyard bully. Often encountering the schoolyard bully becomes a character defining moment that we remember, later on, as important in our development. When we think of expressing Maryness, we usually think about gifts and talents, art, singing, engineering, inventing, cooking, whatever.

To discover, develop, and express those takes freedom. And freedom requires risk. You cannot have riskless freedom. That's an oxymoron. That's the great fallacy of government protection. I find it ironic that these conservative industrial food advocates—look at the voting record and you'll see them leaning Republican—view chickenness in terms that are identical to the suffocating government safety net that they oppose and Democrats endorse. If they lobbied for Maryness policy consistent with chickenness policy, they would want more taxes, more government programs, health care, bigger social security, etc.

Deep down, very few of us wants safety to suffocate freedom. This is the whole problem with the food police. They really don't believe you and I have the capability to make food choices. We might patronize a bad farmer. We might choose unwisely—as if the government's food police choose wisely. The freedom to progress requires the freedom to fail. If we take away the freedom to make an erroneous decision, then nobody can make innovative positive decisions. Living in a strait jacket, in a sealed bubble, being fed intravenously protects us from bruises, cuts, even biting our tongue. But who wants to live that way?

To assign safety as the first and most important aspect of self-expression is both logically ridiculous and intuitively inappropriate. Some of the most distinguishing stories of heroism and self-expression come from life's most tenuous, unsafe, risky moments. Imagine an astronaut describing safety as his number one criterion for self-expression. No, it would be risk, adventure, discovery, going where no one has gone before. That's not a safe place. It's an exhilarating, high risk place.

But beyond that, describing chickenness in terms of safety assumes that you can ultimately have a safe place. Of course, to these guys, the CAFO is assumed to be the safer place. Remember, this is where nine chickens are crammed in a battery 16 inches by 22 inches for their whole lives. These cages, stacked on top of each other, don't even give enough room for the birds to fully extend all their extremities. Confined to these wire mesh cells, the chickens never scratch, never see sunlight, never breathe fresh air. Sounds like a safe life to me. While they are indeed protected from hawks, this living exposes them to a host of other maladies.

The unfair assumption is that such a model is actually safe. In fact, it isn't. Yes, here at Polyface we lose birds sometimes to predators and rough weather. Sometimes the birds gang up on a weak bird and peck her to death then eat her up—starting by disemboweling her and dragging her intestines around. If you ever want to be convinced that humans are not animals, just keep a flock of chickens. They are as perfectly happy to dine on sister's GI tract as corn. Makes no difference.

On the whole, I'm not sure we lose more birds to these pastured production nuances than the industry loses in disease or infrastructure malfunction. When I hear about hundreds of thousands of birds dying in a heat wave, for example, that's in those safe houses, remember? And plenty of sick birds exist in those houses. The whole notion that living in a CAFO is safer than in a field with portable infrastructure and a guard dog begs the question: in which scenario do more birds die prematurely?

I don't think for a minute the pastured model necessarily has more losses. And it certainly provides a happier life for the chickens. We do all we can to protect the birds: field shelters for broiler chicks, guard dogs, electric netting, short grass, periodic grazing cows nearby, hunting, trapping, hygiene. But at the end of the day, neither the industrial system nor the pastured system can deliver 100 percent protection. If there's no guarantee anyway, I'll pick the system that offers scratching, exercise, fresh air, bugs, grass, and sunshine. At least if their lives come to an untimely end, they've been full of discovery and chickenness.

A chef came out to the farm for a visit and we walked up by the pigs in their savannah—pasture and widely spaced trees. Since it was midday, the pigs were all stretched out on their sides, lying in the grooves they'd trenched with their snouts to fully contact cool soil. As we approached, some of the closest ones opened their eyes a bit, blinked at us, yawned, and continued their napping. After awhile, the chef said: "I've never seen live pigs in my whole life. But if I were a pig, this is the way I'd want to live."

The industrialist would counter: "But bears roam here at night. And a cold rain might sweep through tomorrow. And protozoa live in the soil." Yes, that's all true.That's why we don't take the pigs out to savannahs until they are bigger than 75 pounds—so they can fight back. We move them frequently so they have grass and a canopy of trees for protection and hygiene and to break up the protozoa cycle. And usually they enjoy the ecstasy of a day like the one when the chef and I visited them. To confine them on slats with 5,000 other hogs in a smelly, dingy CAFO would deny them this ecstatic, luxurious experience of pigness.

To take this whole discussion one step further, I think it's interesting that the industrialists define chickenness from a standpoint of fear. I define it from a standpoint of intrepidity. Who wants to live life in fear? People who move the world are not the ones who hunker down in fear bunkers. No, indeed. The people you read about in history books are intrepid, running toward the unknown, and being lunatics. I'll take that any day.

With all that said, this whole essence of pig demands that we create a habitat that leverages the plow on the end of the pig's nose. After all, digging, plowing, rooting—these certainly define the essence of pig. That's what separates the pig from a goat or a tomato. On our farm we don't ring the pigs' noses to keep them from rooting. We move them through pig pastures (savannahs) and then turn them into forest glens. In these glens they dig up brambles, weedy shrubs, and briar roots, eating these starchy tubers. They find acorns, hickory nuts, grubs and worms, which they eat heartily. Digging around trees rids the trees of invasive pathogens.

We rotate often in these 3-5 acre glens, generally on a one month on, eleven months rest schedule. By allowing access only once a year, the forest regenerates acorns, weeds, and bugs as a renewed buffet for the pigs. The whole idea is to duplicate the periodic disturbance created by a herd of a million buffalo chased by fire or predators that occurred centuries ago and built these productive eastern silvo-pastures.

In the spring, when the cows come out of the hay feeding shed, where a thick anaerobic bedding pack of wood chips, sawdust, and old hay have absorbed hundreds of tons of manure and urine like a giant carbon diaper, we use the pigs to build compost. As the bedding is building through the winter, we add whole shelled corn along with the carbon bedding. The cows tromp out the oxygen. The entire bedding pack, as deep as 4 feet sometimes, ferments anaerobically. The corn ferments as well.

When the cows come back out onto the pastures at spring greenup, we turn the pigs into the bedding pack. All pigs have a sign across their forehead: "Will Work for Corn." The pigs seek the fermented corn in the bedding and in the process till it all up, injecting oxygen like a big egg beater. Hence the term: PIGAERATORS. Most people are familiar with windrow compost piles, whereby a huge machine turns the material periodically to aerate the pile and stimulate decomposition.

By using the pigs, not only do we save money and machinery, it actually fully honors and respects the pig. This is not a job pigs despise. It is truly hog heaven. If you ever, ever, want to experience happy pigs, come to Polyface during the roughly 60 days each spring when we're pigaerating. It will be an epiphany never forgotten. As soon as the pigs start turning and oxygenating the material, it begins heating.

One of my most enjoyable rituals is to walk out to the barn very early on a nippy late March morning and peek in on the pigaerators. The heating compost steams up through the pigs, who are lying on their sides, flat out on that warm bedding. Enshrouded in that steam, the pigs present a surreal picture, like something out of MacBeth with steaming cauldrons, like a special effects theater. The quiet breathing and occasional yawn or lip-smacking of a sleeping pig on steaming compost—now that's sheer ecstasy.

In similar fashion, we follow the cows with the eggmobiles, which are portable henhouses. The chickens free range out, scratching through cow patties, spreading out the manure, and eating out the fat fly larvae. This not only spreads the nutrients but protects the cows from flies and parasites. The laying hens produce eggs as a byproduct of pasture sanitation.

In both of these models, the animals are no longer just animal protein producers. In other words, the pigs are not just bacon and pork chops. The chickens are not just egg layers. Rather, the pigs and chickens are co-laborers, team players, in this great land healing ministry. What a grand change of emotional relationship with the life on the farm, when rather than seeing it as something that just grows and I sell, instead it is something that works alongside me.

We are all in this together. They appreciate my care and choreography to make sure their dance is appropriate. I appreciate their contribution to the work, the synergy, of the farm. If you attend any industrial farming conference, way more than 50 percent of the lectures and in-hall discussions will be about diseases. Most farmers view their plants and animals as fickle death-wish dependents that the farmer must hover over, worried every minute about the next fiasco.

A friend told me once that he quit raising confinement hogs when he realized that his first waking thought every morning was what disaster he would find at the hog barn. Usually it was a broken slat that allowed a hog to fall down into the manure lagoon, located basement-style under the slatted floor that the pigs stood on. That sounds safe to me—how about you? But you won't hear about this in the industrial pork literature. You won't see all the horrors that led up to that anemic pseudo-pork chop at the meat counter.

I can't leave this discussion without broaching mad cow disease. I don't know what causes it. Mark Purdy, the British dairyman who traveled the world sleuthing this malady discovered convincing evidence that it was caused by a combination of overused organophosphate parasiticides and grubicides in conjunction with heavy metals. He, of course, was considered a lunatic and eventually passed away—some think he was murdered because his findings were becoming too irrefutable.

The official cause, according to government sources, is herbivores eating animal proteins, and especially meat products of the same species. As the processing industry centralized, the concentration of guts became a bigger problem. No longer were they spread out in small quantities that could be composted, buried, and handled appropriately. Instead, mountains of the stuff became a waste problem. What does our culture do with waste problems? We feed it to cows, of course.

For several decades, the USDA wined and dined farmers like me to acquaint us with this new scientific method of feeding cows. The industry took these mountains of cow guts, cooked them, skimmed off the broth and dehydrated the precipitate, then fed that back to the cows. Like feeding dead cows to cows.

Farmers flocked to this new system. All, of course, except the lunatics like me. We looked at this approach and asked: "Is there any place in nature where herbivores eat meat?" The answer was no. It didn't exist. And so the lunatics did not buy what the USDA scientists were selling.

About 40 years later and this big collective "Oops, maybe we shouldn't oughtta done that" exuded from the scientists when bovine spongiform encephalopathy (mad cow) made headlines around the world. Immediately, the industry quit feeding dead cows to cows. Of course, they are still feeding dead chickens and chicken manure to cows, but at least not dead cows to cows.

At the time the USDA scientists were pushing the new dead cow menu to cattle operators, I spoke out against it. I was ridiculed as a Luddite, anti-science, non-progressive, backward barbaric lunatic. After all, this grew cows faster, fatter, bigger, cheaper. Nobody was asking: "Where's the natural pattern? Does this violate any of creation's designs?"

None of the scientists dared to ask such a ridiculous question. Just like the scientist in Jurassic Park, the only thing they cared about was doing it. No moral or ethical or higher questions needed to be answered. Then, in this long period of time between cause and effect, their recklessness with nature's parameters created a deadly consequence. That is why I promote philosophy over science. I'm not opposed to science. But I think as soon as we do not submit science to philosophy, we're heading into troubled water.

Science refused to ask even the most fundamental questions because it is interested not in long term implications, but in today's feasibility. Can it be done, from a technical standpoint, is the only acceptable question. Not should it. But can it. That's the difference between technicians and prophets. Technicians are out here doing things but never asking if they are the wrong things. Allan Nation, editor of *STOCKMAN GRASS FARMER*, likens this to a group of machete-wielding trail blazers hacking through the jungle on their way to a village.

"Wow, look how much we've gotten done," they congratulate each other when they pause to rest. One of them, the prophet (lunatic) offers: "I think I'll climb up in a tree and see where the village is."

"Oh, you Luddite. You naysayer. You anti-progressive. Can't you see how much we've gotten done? Goodness, at this rate we'll be there before nightfall. Who needs to climb a tree? Do you doubt us? Who gave you the right to question? Who died and made you king?"

"Well, I'm going to shinny up the tree anyway and see where we are in the big scheme of things. " So he shinnies up the tree, and it turns out they've hacked a complete arch and are now heading in the opposite direction. "Hey, guys," he yells down, "the village is that way!"

But nobody listens to him because the group of technicians is already busily hacking away, making progress...in the wrong direction. Stan Parsons, founder of Ranching for Profit seminars, used to say it like this: "We've become extremely good at hitting the bull's eye of the wrong target."

When all we have is a culture of technicians, and prophets are called lunatics, I shudder to think how much progress we'll make in the wrong direction. We'll create all sorts of problems that our children and grandchildren can occupy their lives trying to solve. What a wonderful legacy. As an aside, I would suggest that government bailouts of inapprorpiate businesses indicates a technical solution, not a prophetic one.

Since science can't measure all the variables in real time, it ultimately can only be as objective as we can see. If a variable is not in our time period or in our sight (paradigm) we won't know to even test for it. Since scientists turn into vituperative iconoclasts when I say "science is subjective," I'll say instead: "science can never test all the variables. Therefore, it should submit to philosophy—a moral, ethical framework, first. That being satisfied, then proceed."

We could debate all day about whose morality and whose ethics. Clearly, my moral framework of chickenness and the industrial framework of chickenness are diametrically opposed. But can you imagine a group of scientists sitting around wrestling with the issues I brought up in this discussion and agreeing to settle it before proceeding with an experiment? Right now, nobody even considers such a discussion worthwhile, let alone whose morality and whose ethics.

At the end of the day, I welcome any visitor, scientist or otherwise, to visit our farm and then go visit a CAFO with the same species of animal. If animals could smile, ours would be smiling; over there, they are frowning. And that's the sheer ecstasy of being a lunatic farmer.

TAKEAWAY POINTS

1. Animals and plants have specific characteristics that should be respected and honored.

2. Nature has an order that ethically trumps amoral research.

3. Humans are clever enough to create things they can't metabolize.

4. I'm a simpleton.

Chapter 11

Portable Infrastructure

When you think of a farmstead, what do you picture in your mind's eye? Go ahead, imagine a farmstead. Pause.

Now let me describe what you just saw: a red gambrel-roofed barn with attached silo surrounded by white board fence, on top of which is a rooster and inside is a horse. A haystack with pitchfork sticking out of it is at one end of the barn, where a couple of cows, sheep, and pigs lazily munch hay and look out at the pasture. I could go on in this vein, but this picture adorns every children's book involving farms. And it's the favorite logo of industrial food companies.

How fascinating that our dominant farm image centers around a building – as if that's more important than soil, crops, or the farmer.

Do you know what's funny? Every single thing about that picture is wrong. Oh, yes, it's the idyllic farmstead, no doubt about it. But it's not the right picture. Very quickly, here's why.

First, the gambrel-roofed barn. This construction technique creates a nice clear span, but is always built on traditional rock or concrete foundations, sill plate, and wall. Depending on era, the wall can be post and beam type construction or more like studs. At any rate, the sill plates on the foundation are always way too close to the ground to handle deep bedding. We'll talk more about the importance of that in the next chapter. Right now it's enough to know that you can't have deep bedding in there.

The clear span is nice, but it gives no structural strength at ground level to hook on different configurations of gates so different small groups of animals can be sectioned off in handy compartments. With pole construction, this kind of sectional adaptation is real easy because you can always wire some gates up to the poles.

Now the silo. Oh, my goodness. If one icon embodies everything that's wrong in agriculture, I'd say the silo is it. First, it's expensive to build, expensive to maintain, expensive to fill, and expensive to unload. Second, it's dangerous. How many farmers have died or been seriously injured because of silos? Farmer's lung, falling, suffocating—silos have killed and hurt lots of folks. Third, they represent the desperate answer to improper pasture management. Because the pasture is being mismanaged and therefore unproductive, the assumed answer is to plant high volume corn to compensate for low pasture production.

Fourth, they are typically filled with corn or other annuals, which must be tilled, planted, fertilized, and weeded. It's the ultimate high cost feeding system. I call silos bankruptcy tubes. And fifth, silos make anearobic silage that is the number one culprit in acidulating the rumen of cows. Cows want roughage, and this already fermented feed generally plays havoc with bovine digestion. If somebody wants to make silage, they should put it in a bunker at ground level instead of shooting it up in the sky. And it should constitute no more than 10-15 percent of the ration.

Next, the white board fence. First of all, fences around a barn are very hard to keep white. A working corral never has a white fence. You might paint it black, but no real farmer would paint it white. It will be covered in manure and rubbed off in no time.

Then the horse. Ah, yes, the horse. In full disclosure, I'm prejudiced against horses because we had ponies when I was a kid and I got bucked off more times than my sore bottom cares to remember. I'm scared of horses, never learned to work with them, and generally think they are a waste of time. I do watch the Kentucky Derby, though. I just can't figure out how to make a profit with a horse.

If I'd grown up with horses, been trained on how to handle horses, maybe I'd be a little more forgiving. I do appreciate good horses when I encounter them. I'm always in awe of well trained horses, but more in awe of the men and women who handle them expertly. I've always said if they could put a front end loader and power-takeoff on a horse, I'd be ready to consider one. But I'll take my little tractors because that's just me. And a tractor doesn't eat when you turn it off. Of course, it doesn't make little tractors for you, so that's probably a tradeoff. But in this idyllic farmstead snapshot, you're not thinking about that horse pulling a hay mower; you're thinking about riding it for fun. And that can quickly become a profit burner.

I don't hate horses. But I certainly don't think they create a functional farmstead today.

Now the outside haystack, with pitchfork. Any good farmer knows you would never leave your pitchfork outside where the handle will get weathered. And anyone who knows anything about farming knows that a nice big barn like that is for hay storage. Haystacks are what you have when you don't have a barn. If you have a barn, you put the hay inside where it won't weather. What's the point in having a barn if you don't put the hay in it?

Now the rooster on the fence. That's the only part I really don't have a problem with. I like a rooster on the fence. Only thing is, the rooster never gets up on the fence. In real life, he's down chasing the hens around and expressing his roosterness. He doesn't have time to get up on the fence. He crows from the ground. Silly rooster on a fence. Whoever heard of such a thing?

The cows, pigs, and sheep all together in the corral is not very workable. The cows would knock the sheep around. The sheep would eat the pig feed. These are not the most compatible animals in a tight living situation. And besides all that, you want these animals out on pasture, not mucking up the barnlot. This is the reality that most industrial food scientists know when people start talking nostalgically about yesterday's farms. This idyllic scene is actually an incubator for pathogens, noxious odors, and sanitation problems.

When industrial food advocates and scientists pooh-pooh me for wanting to go back, this is exactly what they are thinking about: the days of hog cholera in muddy barnyards, Newcastle disease in filthy chicken houses, Bangs in cows. It's not a pretty picture. This is a favorite tactic of industrial food advocates. And this iconic picture plays right into their hand.

So everything about that idyllic farmstead picture is wrong. Isn't it something how this icon persists? To question it is to question motherhood and baseball. And yet I offer a better approach: portable infrastructure.

At Polyface, portable infrastructure is the signature design that enables us to have all the animals out on pasture. It began in the early 1960s when our family went on a Sunday drive and visited a man who had portable A-frames in a field. I was just a little child and can't remember what was in them: chickens, pigs, calves, goats. I have no clue. But I remember Dad's being animated about the whole thing, realizing that portable shelter gives you lots of options.

The problem with a permanent building is that you can't move it. What if you build it somewhere only to discover that it's in the wrong spot? And how do you rotate animals on clean ground if they have to come to the barn everyday? Or worse, if they live in the barn every day (or house, as the poultry industry prefers to call them). Isn't that a wordsmith coup? We call cow housing barns, and pig housing, barns, but the latest comer to the CAFO concept, poultry producers, decided to call it a house. That sounds clean and sanitary, upscale and contemporary.

When we got back home from the Sunday excursion, Dad began constructing a couple of portable shelters. One was for my brother, Art's, rabbits. It was configured like an airplane. The central cabin was 4 feet X 8 feet and 2 feet high, divided into four quadrants for four does. An 8 foot X 12 foot X 2 foot high chicken wire, light framed run, divideded in half, was affixed to the long edge like a wing. Each side had one of these wings. Each doe could go out in a 4 foot X 12 foot run and eat grass. Once a week we'd move the three pieces to a new spot.

The concept was perfect. But the problem was the rabbits would dig out. Many a day Art and I would wait behind a blind, holding onto a string attached to a propstick holding up a box with a carrot in it. Our rabbit catching escapades are legendary, but we never were able to create a portable housing system that worked. Finally, we abandoned the idea, hoisted the rabbit wing runs up in the rafters of the barn, and left them. A few years later, when my fledgling chicken flock was expanding, Dad suggested we pull those old rabbit runs down, place them back in the field, and put the chickens in them. That, folks, is how pastured poultry a la Polyface got its start—totally serendipitously.

But that's the beauty of lightweight portable infrastructure. It doesn't cost you an arm and a leg so if you need to abandon it, you can do so without emotional or economic suicide. One of the biggest problems in modern agriculture is single use capital intensive highly depreciable infrastructure. What do you do with a confinement dairy operation when a new generation says: "Why don't we just turn the cows out and let them graze, feed themselves, and spread their own manure?"

At such a preposterous suggestion, Dad and Grandpa come toddling out of the house and berate this lunacy: "What? Don't you know we spent our whole lives begging bankers for money to build this, bending rebar, pouring concrete, and putting this together? How dare you even suggest walking away from it?" The economic and emotional enslavement that such infrastructure creates in a farm family is enough to drive innovative children away. Change is just not worth the fight.

And even if the farm abandons that structure, it's hard to retrofit for another use. This is why these big old iconic barns all over the country are falling down. They were built for the convenience of humans, not for their animal friendly functionality. Anyone who knows me knows I despise old bank barns. These are barns built into a hillside in order to have two ground-level entrances. Yes, they are pretty, but they show how long people have been thinking about their own comforts instead of the comforts of the livestock.

The reason for bank barns was for ease of feed flow. Building into a bank gave second floor access for grain, hay, fodder, and straw storage. With all the livestock below, on the first level, all these feedstuffs could be shoveled down into mangers and bunks rather than toted horizontally or shoveled up. Only one problem: that basement was a veritable death chamber for livestock. First, it was dark. Pathogens love darkness. Second, it was damp. Pathogens love dampness. Basements are damp. Third, it was stuffy—not enough air flow. That encouraged respiratory problems.

I like big old barns. They are marvelous architectural achievements. But I hate seeing animals in them. They are great for Bed and Breakfasts, square dance halls, restaurants, miniature golf courses and on-farm sales buildings. Anything except livestock shelters. Our farm had a big bank barn, but when a wealthy out-of-state couple bought the place in the 1940s and salvaged the old house, they thought the barn detracted from the house so they tore it down. I'm so thankful they did, because if they hadn't, we'd probably still be trying to make it work.

In addition to being unfriendly to livestock, the ground floors are too low and too tight to maneuver a farm tractor. Skidsteers can get in, but lots of farmers don't have skidsteers. The cleanout is problematic, unless you have lots of cheap labor. I advise people who have bank barns to tear them down, salvage the lumber, and build a pole structure twice the size. I know, to abandon that nostalgic, iconic feature of America's rural landscape is considered lunacy, but I'll take the sheer ecstasy of an animal friendly, adaptable, easy-to-maintain pole structure any time of day.

In the 1950s, after another couple bought our farm, they erected the pole barn that is still here. We've added onto it a lot, but that core structure is still called "the old barn." Unfortunately, they located it in a low area. We've spent a lot of money excavating, ditching, hauling in big rocks out of the mountain, and gradually getting the area to drain. It's still not perfect, but it's pretty good. Probably for the money and time we've spent dealing with drainage issues, we could have erected the barn on better ground.

But it was there. It just seems inefficient to move it. So like the typical farmer, we just try to live with it. Permanent buildings need to be cited carefully. This is one of the biggest mistakes new farmers make—inappropriately citing buildings. Then they're stuck with it for the rest of their lives. And usually the problems extend to the next generation.

I think the Native Americans at Powhatan village near Williamsburg had the right idea. They built out of bent over saplings and the whole house composted in about 10 years. If you think about it, 10 years is about how long your housing needs remain static. As newlyweds, all you need is a tiny place. Kids come along and get bigger, you need a bigger place. Kids leave, you need a smaller place. And those are about 10 year increments. I like the idea of a compostable house. In the long run, it's probably a lot cheaper than building one that will last a hundred years. Every time you build the new compostable house, you can incorporate all the new technology for solar, energy, and waste efficiency.

Portable infrastructure, then, is lightweight and cheap enough that you have the freedom to abandon it, redesign it, or situate it in a different place. After the rabbit run, which morphed into the pastured broiler shelter, Dad built a portable veal calf barn. It had a stanchion on each corner to tie up a milk cow. A trailer about 12 feet X 14 feet, it had an expanded metal floor and was divided into four quadrants. A pair of calves stayed in each quadrant and their manure went onto the pasture through the floor.

At feeding time, we'd bring the cows up to the veal trailer, fasten the first one to a corner, open the calf door, and let the calves out to nurse. Then we'd do the second cow, third, and fourth. By the time the fourth cow was being nursed by her pair of calves, cow number one was about finished. Then we let each cow go back to grazing in sequence. The veal trailer worked, but we finally abandoned it when Dad's accounting job upset the milking routine and the calves suffered due to the sporadic schedule. My older brother Art, younger sister Loretta, and I were not yet big enough to take care of this chore.

As discussed in chapter two, Dad developed the portable electric fence. Over the years, as internal board or woven wire fence deteriorated, we did not replace it. I'm grateful that we didn't, because as time went on and electric fence technology improved, we realized there was no reason to have internal permanent high cost physical fence. As a result, we've been able to fence appropriately to the lay of the land and functionality of the livestock.

Every time we lease another farm, I'm struck by how many costly high maintenance fences are in the wrong places. You know, those straight fences I talked about in chapter four. These fences are as hard for a farmer to abandon as a barn. So even when they are in the wrong place, farmers try to keep using them because they are such a defining feature of the landscape. One of my rules of thumb for fence building is this: don't put in any permanent fence for three years. Make everything temporary. Whatever you haven't moved in three years, convert to permanent with better posts and stronger electric fence wire. That way you let function drive form, instead of the other way around.

When I began looking at portable laying hen houses, I built a 6 foot X 8 foot peaked structure roughly 2 feet high at the eaves and 3 feet high at the peak, on bicycle wheels. It had a light mesh floor and I could just push it around. I used two sets of triplicate lightweight poultry netting gates to set up a hexagonal yard. Every couple of days I'd spin the house 90 degrees and rotate the hexagon to the adjacent spot. In a week, I'd cover the entire circle and move the little eggmobile to another area and repeat the circular process.

That system worked so well that I decided to retrofit the little prototype to a 3-point tractor hitch and move it behind the cows without the hexagonal grazing paddocks. Letting the birds totally free range enabled them to practically live off the land, eliminating feed costs. The next year I built the first 12 foot X 20 foot eggmobile and now we have several eggmobiles.

Since we put these layers in hoophouses in the winter, we can have these eggmobiles out on leased farms in the spring, summer and fall, then just tow them home in the early winter to put the hens back in the hoophouses. It's a simple and efficient way to move the birds around to appropriate housing depending on the season. When it came time to build winter housing, we erected simple and cheap hoophouses (tall tunnels). Layering poultry netting inside the first four feet protects the plastic from pecking and scratching chickens.

When the chickens go back outside in the spring, we plant vegetables in the hoophouses. This is multi-use infrastructure. If you're going to build a structure, never build it with only one use in mind. More about that next chapter.

The one drawback to the eggmobiles is that they are land extensive. If you park it too close to the house or flower gardens, guess where the chickens go. And guess what the flowers look like.

Using high tech lightweight poultry netting from Premier (polyethylene webbing interwoven with stainless steel conductive threads and fiberglass push-in stakes affixed every 8 feet) we developed what we call the Feathernet system. This is land intensive because the birds are contained inside the electrified netting. Not only does the netting keep the birds in, it also keeps predators out.

A scissor-truss A-frame looking contraption on skids is the newest design. At 20 feet X 32 feet, this structure works great for 1,000 layers. A quarter acre paddock created with the highly portable netting is plenty of pasture for 1,000 birds for three days. We move the whole contraption every three days by erecting an adjacent netting circle, opening both circles up where they meet, and then pulling the shelter through to the next paddock. The layers just walk into the next paddock with the shelter. This allows us to move these birds being contained in the netting right up around the flower gardens. This is a land intensive system, especially good for smaller acreages.

While this may sound cumbersome, consider the following:

1. No concrete.
2. No rebar.
3. No building inspection.
4. No taxes—it's not a building and it's not a farm machine. Nobody knows what it is, so it's completely untaxable.
5. Expense it all up front. No need for depreciation.
6. One-third the size structure required for the same amount of loose housed birds because they spend so little time under the cover and most of their time out in the pasture. This is just a bedroom and place to lay eggs.
7. No grading or site preparation.
8. No bedding to haul in to absorb the manure.
9. No manure to haul and spread.
10. Much happier chickens.
11. More nutritious eggs.

When you add it all up, the portable shelter has a lot going for it. And the light footprint makes a more beautiful landscape. The dominant features of the landscape too often are the monolithic structures people build, instead of the natural beauty of the landscape itself. That's one of my favorite things about electric fences. Since their strength is invisible electricity, their physical presence is almost invisible. That makes a much more pleasing landscape.

Now to turkeys. We use the same electrified poultry netting to contain them, moving them every couple of days. But since they are not laying and enjoy roosting, we made a Roostmobile. This is a scissor truss on an extended hay wagon chassis. A peaked roof creates a lot of internal space, which we filled up with perch boards. The turkeys can hop up on the lowest perch and gradually work their way up. Once the top perches fill, the lower ones begin to fill, but all of them are too high for most predators to reach. Since this is only a bedroom, a relatively small Roostmobile, say 12 feet X 20 feet, is plenty of shelter for 400 turkeys.

The Roostmobile enjoys all the same benefits as the Feathernet.

When the rabbits are weaned, they go out into Harepens, which are about 30 inches wide, 6 feet long, 2 feet high, with a slatted floor. Floor slats keep them from being able to dig out, but offer 70 percent of the ground open for grazing between them. Extremely light, these little Harepens weigh only about 30 pounds and a child can move them by just grasping the handle and sliding them along. They slide on the floor slats. The rabbits just ride along as if they're on skis.

Just to show how versatile portable infrastructure can be, several years ago we tried sheep. For housing, corral, and everything, we built a 30 foot X 48 foot hoop structure called the Ewego that we pulled around. We used the electric netting to create paddocks radiating out from the hoop structure, like leaves of a four-leaf clover. After the four paddocks were grazed, we'd move the whole structure and do another four-leaf clover. We only needed enough fence for one paddock because we'd just put the sheep in the Ewego while we took down and moved the grazing paddock.

After three years we abandoned the sheep. We had gotten them hoping they would graze weeds that the cows didn't like, but since our pastures are so clean, the sheep competed with the cows rather than complementing them. And since they took more time per dollar in sales, we abandoned them. That summer too Daniel was trying to finish his house and we had a difficult apprentice situation so something had to go. Since sheep were the last thing in, they were the first thing to jettison.

But the story is not all sad. At that time we needed a new Raken (Rabbit-Chicken) house because we were outgrowing the one we'd had for 15 years. What to do? We just pushed fence posts in the ground on 4 foot centers, affixed boards to the inside up about 4 1/2 feet, and set the Ewego up on top of that pony wall. Voila. A new Raken house. In the winter, we move the layers and rabbits out of the Raken and into the hoophouses. Pigs go into the Raken house. We switch them back in the spring. Multiple use and easily adaptable.

Another wonderful aspect of portable infrastructure is that it's ideal to use on leased land. Landlords can come and go. Because land control can be tenuous, having portable infrastructure offers complete versatility in location. If I change land bases, I just tow everything up the road to the new location and I'm right back in business. I don't have to worry about investing in infrastructure on rented land. That's a huge benefit of portable infrastructure.

We use guard dogs with our pastured poultry. We train the dogs to electric fence with a Cabela's buzz collar. Because a dog can only protect about 20 acres, we move the dogs around every few days. They leave a protective halo because it takes a few days before the predators realize the dog is gone. By moving the dog onto another 20 acre field every few days, it can actually cover 60 acres.

Some grass farmers spurn portable infrastructure in favor of a permanent building surrounded by rotated pastures. I realize I may upset some farmers by saying this, but I don't see how a permanent structure with rotated pastures can be sustainable. It may work even for a few years. But if you do the nutrient analysis, the toxicity load, and the pathogen cycle, I think you'll find that gradually the grass paddocks will falter.

One caveat: use very low stock densities and multiple species. For example, sheep rotate through twice to every time the chickens use the paddock once. Or use a milk cow in conjunction with the poultry. Another option is to spread sawdust or wood chips on the paddock in order to give the nitrogenous poultry droppings something carbonaceous to digest. But by the time you do all the necessary management to make such a system work in the long haul, my sense is that you might just as well go to portable infrastructure and not have the higher maintenance at best, and land injury at worst.

Clearly, modern American agriculture exults in big buildings, lots of concrete, lots of rebar, and big landscape footprints. I remember well many years ago being the counterpoint of a presentation by an agriculture historian from the Smithsonian. He pressed the point, over and over, that bigger is always better. He said that a bigger fort was always better than a smaller fort. So a bigger corn plant defeats a smaller corn plant. A bigger turkey defeats a smaller turkey. Over and over, he said bigger is better. He saw that as the best argument against environmental agriculture.

A bigger tractor is better. A bigger pig barn is better. A bigger chicken house is better. The only problem is that when you take that position, at some point you will exceed the point of efficiency. Joel Arthur Barker's classic book, *PARADIGMS*, which popularized the word in American business culture, points out the axiom that every paradigm eventually exceeds its point of efficiency.

That's why the notion that companies can become too big to fail, and therefore merit governmental bail outs, is devastating to innovation. If a paradigm cannot go through its size-limiting collapse, the replacement innovative paradigm cannot take root and grow. Large corporate safety nets retard innovation and creativity.

I remember well when the first French Concorde, the SST, landed in John F. Kennedy Airport in New York. It headlined all the newspapers. Big front page full color pictures. That sleek droop-nosed aircraft wowed the world. All the pundits predicted it was the harbinger of things to come. The next generation would be bigger, faster, sleeker. This was just the beginning.

Who would have thought that within 30 years, not only would the next generation not arrive, but the SST itself would be obsolete? It was too big, too fast, too expensive. It was replaced by slower, stodgier aircraft. But aircraft with more balanced costs per mile. Barker was right: every paradigm eventually exceeds its point of efficiency.

That's why we need to stay portable in our thinking as well. The one-agenda policy wonks, whether they be drill-for-more-oil or lock-up-more-wilderness zealots, are usually lopsided. Their thinking, when run to its conclusion, will topple. That's why we need to be widely read. It's good to listen to conservative talk radio at the same time you're reading the Huffington Post. Move your mind around; engage people you don't agree with. That helps you move to a more balanced approach.

Portable infrastructure, while it may not have the wow factor of a gargantuan monument, does not dominate the landscape. It preserves more freedom, more options for future adaptation and the people running the farm. It's the only way to really keep the animals on clean, fresh ground. All in all, it's key to the sheer ecstasy of being a lunatic farmer.

TAKEAWAY POINTS

1. The iconic red barn, silo, and horse illustrate wrong-headed agriculture.

2. Portable infrastructure enables modern pasture-based operations to maintain sanitation and protection.

3. Portable infrastructure preserves more options for the farmer to either innovate or use someone else's land.

1. Eggmobiles follow the cows like birds follow herbivores. Laying hens scratch through the dung to eat parasites and fly larvae, and spread the dung like a huge biological pasture sanitizer. In addition, the hens eat grasshoppers, crickets, worms and other protein-rich animals, converting them into nutrient-dense dark-yolk eggs.
Photo by Rachel Salatin.

2. Pigaerators seek fermented corn distributed through fermenting, anaerobic cattle bedding in the hay shed. Carbonaceous material acts like a giant diaper to hold manure and urine until it can be oxygenated and aerobically composted with driverless, appreciating, petroleum-less pigaerators. Hog heaven indeed. *Photo by Rachel Salatin.*

3. Meat chickens called broilers move across the pasture in floorless pasture schooners, receiving a new salad bar each day. We use a simple dolly, sitting to the side, to lift one end of the schooner and pull it to the next pasture buffet. The chickens walk as their house moves. The shelters protect the juvenile birds from predators and weather. In addition to a new salad bar, the birds receive a fresh lounge area each day. Kind of like having their bed linens changed every day.
Photo by Rachel Salatin.

4. A real stump speech—delivered by a blueblood turkey. A turkey roostmobile provides shelter for the turkeys as they go about expressing their turkey-ness. Who knows how many elections and bail out plans they orchestrate as they commune through the pasture preparing for November elections? Are you ready to vote? *Photo by Rachel Salatin.*

5. Mob stocking herbivorous solar conversion lignified carbon sequestration fertilization. These 350 head of cattle have just converted tons of biomass into nutrient-dense salad bar beef and are now rushing into their next paddock. Long rest periods and 24-hour paddock stays stimulate vegetation to collect more solar energy and pulse more carbon into the soil. If every cattle operation in America practiced this technique, all the atmospheric carbon created during the industrial age would be sequestered in fewer than 10 years. All fertile soils around the world have been built with perennials, movement, and herbivores. This bio-mimicry is still the most efficacious planet-healing model available.
Photo by Rachel Salatin.

6. Acorn finishing glens leverage traditional pig-forest symbiosis. Modern electric fencing and simple nylon rope insulators enable ecological massage using highly controlled pigs to work their magic. The periodic disturbance freshens up the forest floor to germinate more diversified species, incorporates leaf litter to enhance soil building, and protects trees from borers and grubs. *Photo by Rachel Salatin.*

7. The millennium feathernet is a pasture-based intensive egg production system that uses space-age polyethylene webs entwined with stainless steel filaments to carry pulsing electricity that keeps predators out and poultry in. For the first time in human history, marrying techno-glitzy to the chicken-ness of the chicken, we can raise large flocks more hygienically, ecologically, and sanitarily than a backyard flock of chickens on an American homestead circa 1900. Is that cool, or what? The scissor-trussed portable shelter surrounded by electrified netting is home to 1,000 egg layers who move every three days to a new pasture paddock.
Photo by Rachel Salatin.

8. Open-air communal chicken processing. Welcome to the world's cleanest chicken, and the system demonized and criminalized by American food police. Fresh air, sunshine, and only processing a few days per month, seasonally, insure far safer birds than available through government-approved centralized factories. Kill, scald, pick, eviscerate, quality control, chill. Efficient, but an enjoyable place to work because it's nested into the landscape where birds sing and flowers grow.
Photo by Rachel Salatin.

9. Harepen for rabbits. Portable, slatted-floor (to keep the bunnies from digging out) shelters provide fresh pasture and clean lounge areas for meat rabbits. For more than 20 years these vaccine-and medication-free line-bred rabbits have adapted to a forage-based model. Preferred by the best chefs, these rabbits have a hare-raising life.
Photo by Rachel Salatin.

10. Raken house (*Ra*bbit-Chi*cken*) illustrates symbiotic, relationship-oriented farming at its best. The rabbits above drop urine, manure, and food left-overs that the chickens underneath scavenge or scratch-incorporate into the bedding. Plenty of carbon creates vibrant decomposition, which in turn germinates seeds and stimulates worms and bugs to grow. Nematodes keep pathogens in check. You could eat lunch in here.
Photo by Rachel Salatin.

Chapter 12

Pathogen Cul-de-sacs

Louis Pasteur is the famous French scientist who advanced the germ theory, a world of bad guys out there trying to get the good guys, which were victims. His whole idea was to kill all those bad guys before they hurt the victims.

What many people don't realize is that Pasteur had a contemporary named Michael (pronounced Michelle) Beauchamp who argued "Au contraire!" Beauchamp agreed that the bad guys were out there, but advanced the theory that the bad guys could only win if the protective defenses allowed them access to the victim. This has commonly been called the terrain theory. Beauchamp said sickness was all about the terrain.

Terrain encompasses many things: hygiene, stress, immune response. The two scientists argued throughout their careers, but Pasteur was handsome, flamboyant, and did better Good Morning America interviews. Besides, Pasteur's germ theory was more acceptable because all of us would rather be victims. Beauchamp's terrain idea meant the responsibility was ours to create an immunological terrain to keep the bad guys at bay.

Interestingly, on his deathbed, Pasteur rose on an elbow during a brief time of lucidity and moaned: "Beauchamp was right. It is all about the terrain." Then he fell back onto the bed and expired. It's one of the most famous recantations in all history.

The important part of this story is that the western world bought Pasteur's theory but did not buy Beauchamp's. Most people have never heard of Beauchamp. And yet Beauchamp's theory launched the entire wellness movement. It's the whole basis of the heal thyself paradigm, naturopathy, homeopathy, acupuncture, chiropractic—basically, everything western medicine calls quackery. And most Americans have never heard of him. They don't even realize that an equally valid alternative sickness paradigm exists. Such is the nature of truth. Mob endorsement is seldom healthy. You'll always find the truth around the fringes, lurking in the forgotten corners, the unheralded buried news.

As a matter of full disclosure, and I'm sure as a complete surprise to you, let me say that I am an absolute believer in the terrain theory. I'll start by telling two stories. Many people ask me: "Have you ever had an animal sickness on your farm?" If they don't ask that question, they ask this one: "What do you do when you have sickness on your farm?"

First of all, we certainly have animals get sick from time to time. Anyone who grows things deals with sickness. If it's a chicken, we don't do much. A chicken isn't worth a vet bill. About the only thing we call a vet for is a calving difficulty too hard for us to handle. If we buy a calf and it gets sick during the new farm transition, we may administer an antibiotic. But mainly we segregate poor doers, take the herd pressure off them, and usually they straighten right up. The fact is that when you have the tens of thousands of animals that we do, you lose one once in awhile. Goodness, in a group of people that big you lose a few too.

In the 50 years of our farm's existence, we've had three disease outbreaks. The first was many years ago with chickens. A lady asked us to raise 500 pullets for her because she wanted to get into the egg business, but didn't want to raise the little chicks up to 5 months of age, when they begin laying. We purchased the extra chicks and brooded them with ours. As spring approached, when they were big enough to put out in the field, the weather stayed cold and rainy. In fact, we ended up calling that season the year without a summer.

Our normal rainfall is 31 inches a year, and we had 50 inches by March 15. Everything was soggy, and cold. We waited for things to warm and clear. Meanwhile, the chicks were growing and getting crowded in the brooder. Really crowded. Finally, we could wait no longer because our first batch of broiler chicks was scheduled to arrive and we needed the brooder for them. So we reluctantly went to the pasture with these roughly 10 week old pullets. The weather remained terrible. Cold and wet. Day after day. The birds started dying.

We began losing 10 a day. They would just begin to shrivel up. It took about 48 hours from onset of droopiness to death. We ended up losing nearly 50 percent of that whole batch. Oh, by the way, the lady decided she didn't want to go into the egg business after all. Thanks a lot. I started looking through my *Merck Veterinary Manual* and other poultry books and soon pinpointed the problem: Marek's disease. No question this was what we had. It was a dead ringer.

Description: caused by unsanitary, unhygienic, stressed conditions. Hits birds at about 8 weeks of age. Finally runs its course and you have left what you have left. Can kill up to 80 percent of a flock. Cure: vaccinate. In the industry, most birds routinely receive the Marek's vaccination as a matter of course. We had never vaccinated because we never had a problem. We've not had a problem with Marek's since. We did vaccinate for a couple of years after that because it lives in the soil for several years, the books said. But the main point of the story is that we caused the problem due to overcrowding and stressful conditions. That's terrain, and it was completely our fault.

Story number two. Several years after beginning to raise the pastured broilers, we began having a lot of trouble with the chicks not being able to walk. Their toes would curl under and they couldn't keep their balance. Finally, they would become completely immobile and then die. Again I went to my books and found it soon enough: curly toe syndrome. In fact, the problem scared me enough that I did something I wouldn't do today, as distrustful as I have become of the government. I actually took a couple down to the state lab for diagnosis.

What happens is that in these fast-growing meat birds, the nerves, which grow slower than the skeleton and muscle, don't keep up with the growth of the nerve sheath. The friction sets up a staph infection in the nerve sheath. This begins the paralysis down in the foot, curls up the toes, and eventually kills the bird. The state lab gave me a whole sheet listing antibiotics that I could administer. Interestingly, more than half were marked as unusable because the industry had used those antibiotics so much that the staph had built up an immunity to many of the drugs.

I refused to use the antibiotics, so I headed into my wellness books. Vitamin B is the nerve vitamin, so I reasoned whatever would be high in Vitamin B might help my birds. Liver and leafy greens. Bingo. That, I reasoned, was why after the chicks were on pasture a couple of weeks the problem just disappeared. The curly toe was hitting them at about day 7, in the brooder. They normally went to the field sometime between 14 and 20 days. I began feeding liver and that stopped the problem immediately—even in the birds that were already practically immobile. The toes straightened right out, the birds began walking, and it was a complete heal.

That means the nutritional solution actually stopped the staph infection. How about that, all you Pasteur disciples? Eventually, we fortified the ration with B vitamins by using the Fertrell Nutri-Balancer feed supplement and we've never had a problem since. But again, that was a ration deficiency that was our fault. You could argue that it was a genetic defect created by selecting for supersonic growth characteristics. And certainly that's true. We've never had a problem in the slower growing non-hybrids that we use for layers. But still, once we got the ration right, the birds were and have been fine, industrial genetics or not.

Third story. We leased a nearby farm that had been mismanaged—like all of them have been—and it had huge briar thickets. I'm talking two acres of solid blackberries 12 feet tall. Br'er Rabbit heaven. The landlord did not want us to mow the thickets because he wanted to be able to pick blackberries. We didn't want the blackberries because we were renting the land by the acre and the cattle couldn't even penetrate these thorny thickets.

When we initially took the herd over to that farm, we put the mineral box in these thickets to encourage the cattle to tromp down the brambles. Tromping is not mowing. Heh. Heh. Over time, we reasoned that the combination of hoof action and manure would eradicate the brambles. But we had never had thickets of this magnitude. They were legendary.

After a week there, we lost a calf. Not a poor calf, but a really good calf. The next day we lost two more. The next day we lost two more. These were nice 500 pound stockers. Worried to death, I called the vet. He came and autopsied one. Took him about 5 minutes to make the diagnosis: blackleg. Around here, every farmer vaccinates for blackleg. I remember losing one or two calves to blacklag back in the 1970s, but had never seen it since and had forgotten what it looked like.

It attacks the nerves, turns the muscles dark (hence the black) and attacks the most rapidly-growing calves. They are the ones with the fast metabolism that allows the protozoa to invade the system quickest. According to the vet, the only cure was to vaccinate. The next day, we put the whole herd through the headgate and vaccinated. We lost one more, but other than that, they did fine.

But I still wasn't satisfied. I had weaned myself philosophically from the mandatory vaccination paradigm and I sure didn't want to admit defeat. So back to my bookshelf I went. One of my hobbies is collecting old farm books. I love material written before 1940. It's full of wisdom. For example, any pre-1940 swine book advocates feeding pigs charcoal as the number one hedge against sickness. We have taken pigs that were as good as dead, let them eat charcoal, and in a couple of weeks they are as healthy and slick as can be.

Anyway, I looked up blackleg and read everything I could. Everything was negative. No cure. Terribly communicable. If your cows get it, you're doomed. Disposal: dig a pit, burn the carcass, fill with lime, cover over with dirt, don't graze that area for 200 years. Oh, it looked bad. I finally reached the last book in my cattle repertoire, and admit that I was in despair. But the last sentence of the last paragraph, said this, in essence (some of these disease texts get pretty technical): "Protozoa gain entrance through anaerobic puncture wounds, often encouraged by brambles."

Eureka! There it was. We'd turned these calves into veritable anaerobic pin cushions. No wonder we had blackleg. We'd been enticing them into these bramble thickets with the mineral box. Immediately we received permission from the landlord to declare Jihad on the brambles, and have never had another incident since. And we don't vaccinate for blackleg. And we didn't lime and we graze right over where the buzzards cleaned up the dead carcasses.

Again, just like the other incidents, this was completely our fault. So in my lifetime, I've gone through what I would call epizootics three times, and every single time it was undeniably and completely my fault. Beauchamp would call this the terrain. You see, as soon as I take responsibility for the problem, the answers appear in a management or nutritional protocol, not in a pharmaceutical bottle. I can assure you that this personal responsibility approach is considered lunacy by mainline agriculturalists.

Perhaps the most significant difference between sustainable agriculture conferences and industrial agriculture conferences is the amount of time devoted to diseases and sickness. Perhaps 10 percent of a sustainable conference agenda will be devoted to diseases and health problems. But about 90 percent of industrial food conference agendas deal with pathogenicity. Why the difference? It's all about the terrain. It's all about the terrain.

Rather than always circling the wagons and shooting at the bad guys, our side is out here building a community the bad guys don't want to join. Our side is propelled by optimism, systemic solutions, prevention, and optimal health. The germ side is propelled by pessimism, bandaid coverings, diagnosis, and fear. The germ response is always after the fact. The germ theory is about killing bad guys faster than they can kill you. The terrain is about having such a wholesome community shindig the bad guys think you're nuts and go somewhere else.

When I speak at conferences, the one livestock farming area I feel very ignorant about is disease. During Q/A sessions people often ask about specific problems they are having with their chickens or pigs or cows or potatoes. Often I just have to punt: "I'm sorry. I just don't know much about diseases." And I usually admonish the questioners to look at their protocol and ask themselves:

1. Do I have any bare dirt under my animals?
2. Are my animals ingesting all the green grass they could? Am I optimizing the quality and conditions of the grass so that ingestion is as high as possible?
3. Am I confusing the pathogens by keeping them offguard with symbiotic diversification?
4. Am I creating host-free rest periods routinely?
5. Do I have a vibrant decomposition process ongoing?
6. Is the drinking water clean?
7. Am I masking weaknesses with a vaccination program?

These are all valid questions, and speak directly to terrain. If we routinely vaccinate or medicate a group, how do we know the genetic propensity to sickness? Back in the mid 1970s on our farm we quit using a grubicide to control heel fly warbles. This is a parasitic fly that in the spring lays eggs on the hair follicles of the lower hind leg—right above the hoof of a calf. The egg hatches out and a tiny worm crawls down the hair, burrows into the skin, and then burrows slowly up through the animal until it pops out on the back.

It secretes a saliva that dissolves the hide so the now inch-long grub can breathe. When it's mature, it wiggles out of the hole, drops on the ground, pupates into a heel fly, and the cycle starts over. Every farmer I know routinely pours a systemic grubicide onto calves in order to kill these parasites. That's just acceptable practice. Anyone who doesn't participate in this eradication program is guilty of animal abuse, according to conventional thinking.

I'm guilty. You can feel these warbles easily in the late winter once they approach maturity. By rubbing a calf's back, you can feel the bumps and count them. I found that we had some calves infested with 15 of these warbles, but others with just 5. I decided to use this as a cow culling criterion. Late every winter, we would put the calves through the chute and physically squeeze out these pesky parasites. We made a note of which calves had the most and least, and in the fall culled the two cows whose calves were the most heavily infested.

Within just a couple of years, we had dropped our average infestation 30 percent and even had a couple of calves that were warble free. My problem with farmwide prevention using subtherapeutic pharmaceuticals is that you can never discover genetic weaknesses. If everyone in a room walks in on crutches, some being okay and others being injured, the quickest way to separate the injured ones is to get rid of the crutches. Then suddenly the strong ones become apparent. That thinking, however, is completely opposite mainline American agriculture. Everybody—and I mean everybody—thinks health comes in a bottle. It's all about the germ theory.

Everybody else, when I speak against patronizing these jugs of things that end in –cide, the Latin suffix for death, thinks I'm absolutely loony. They live in abject fear that the bogeyman is going to kill their cows or chickens. If you listened to them talk, you'd think they actually believe there are evil fairies hovering over their farms, ever so often dropping pathogens out of the sky to afflict their animals or plants with some disease. No rhyme nor reason. No explanation. Just happenstance. "Oh, woe is me for getting picked to be dumped on today. I'm the victim today. Don't laugh, you may be the victim tomorrow." And if I suggest that they could do something about it, I'm unfeeling, uncaring, uncompassionate, and wretched. A terrible neighbor. Aargh!

If our culture tried to design a system that encouraged pathogencity, CAFOs and single species production could hardly be surpassed as an efficacious protocol. We couldn't build a more pathogen encouragning model. Where do you see a template in nature for acres and acres of a single plant? Nowhere. The very notion assaults natural normalcy. And yet in our modern collective agricultural mentality, we don't even question these practices because the unspoken assumption is that they are efficient. And that if we abandon them for more natural templates, the whole world will starve. Here again, we've allowed this sudden infusion of technology and cheap energy to create an illusion of efficiency. If the same technology and energy had been applied to more natural templates, we'd be spinning circles, measured in both health and productivity, around the current crowded single-species model.

Instead of depending on a concoction in a bottle, on our farm the animals and plants are healthy by default, because we create a pathogen-limiting terrain.

Let's go over the essentials. Nature sanitizes primarily two ways: rest and sunshine, or vibrant decomposition. Let's take these one at a time. The rest and sunshine model is what we find in grazing systems and crop rotations. Nothing sanitizes like sunshine. As soon as you put a roof over something, the pathogens have a freer hand to proliferate. That's one reason all the buildings at Polyface have skylights in them. We want that sunshine beaming right onto the floor.

Rest defines a host-free period. The whole idea of rest is to break the pathogen-host feeding cycle and make the pathogens work to survive. In his wonderful book *BIRD FLU: A VIRUS OF OUR OWN HATCHING*, Michael Greger quotes the World Health Organization's Asian director: "outbreaks of avian influenza correspond to where [poultry] population density is very high." It doesn't take a rocket scientist to understand, empirically or intuitively, that the first step in breaking pathogen virulence is to withdraw the host.

That gives the good guys, who are always present (unless they've been nuked by broad spectrum sanitizers) time to attack and begin winning. If a school district has a terrible outbreak of flu, for example, health officials don't round all the kiddos up and confine them in the school gymnasium for a week until the bug passes. No, they close schools. They tell people to stay home. Don't even visit each other. The whole plan is to reduce host congregation and starve out the flu bug

In a CAFO, the animals are always overcrowded (congregating) and the hosts are always present. As Greger points out, the pathogens don't have to spend any time trying to survive— eating, mating, reproducing. They have the luxury of focusing all their energy on becoming more virulent, more adaptive to fumigants, sanitizers, and drugs. The result is superbugs, which the medical community is now recognizing more than ever. If anyone thinks clever human concoctions can stay ahead of these bugs' ability to mutate, adapt, and build protective genetic memory, I have a bridge in Brooklyn to sell you.

In the microbial world, everything happens fast. We're talking about generational transfer every 48 hours. That's fast. And yet the industry moves cavalierly along, building bigger confinement operations, concentrating more single species in bigger fields or under bigger roofs, assuming that tomorrow's scientists will always have an answer. Folks, that's insane. But they call me a lunatic for my lack of faith.

The rule of thumb is that if we deny a host for at least two 21-day periods a year, that's a big enough disruption to keep pathogens in check. Not annihilate them, but at least keep them under control. Can you imagine a CAFO liquidating its stock for two 21 day-periods twice a year? Absolutely not. Beyond that, can you imagine a CAFO putting a different species in the buildings for half a year? That's ludicrous. Structures built for pigs can't handle cows. Structures built for chickens can't handle pigs. And picking and packing equipment built for tomatoes can't be used for squash.

Just to be sure that my broad brush stroke is wide enough, let me point out that this same need for rest periods exists for pets and horses. A dog kennel, or dog run, needs to be vacated routinely as well. And that horse stable, or horse stall—yes, that pet horse on that overgrazed paddock out behind your house—that's a pathogenic playground. Nature creates this mandatory rest period a couple of ways. One is by moving animals on to new ground, via grazing rotations.

Grazing rotations occur due to seasonal forage growth cycles, like following rains, as well as repugnant grazing following fire or drought. Animals are pushed forward both by predators and the fly cycle, which is about four days. A permutation of this idea is low density. The reason you see one rabbit here and one over there is because they can't handle high densities. Deer are that way. When they become crowded, they become stunted and sickly—and easier for coyotes and wolves to catch.

The other way nature insures host-free periods is simply by diversifying the species. This keeps pathogens confused. Most pathogens don't cross-speciate. When multiple species run together, or in proximity, or in rotation, it keeps the pathogens off guard and breaks easy living. Do you realize this microbial drama is occurring over us, around us, under us, in us every day? I get a kick out of just fantasizing about all this activity occurring around me. Like the Peanuts character pigpen, surrounded by a cloud of dirt. You and I are surrounded by this community of microbial activity.

Nature has lots of safeguards to protect us from that cloud infecting us. We have immune systems, soap, water, nutrition...lots of things. Our responsibility is to create a terrain that allows us to co-exist without being consumed prematurely.

All this being said, at Polyface we go to great lengths to make sure host-free periods exist. The Raken house is home to rabbits and chickens in mild parts of the year, then pigs in the winter. The winter hoophouses that are home to rabbits, chickens, and pigs are vacated in the spring and we plant vegetables in the bedding. The hay shed that houses cows in the winter houses pigs during pigaerating in the spring, and then can house chickens in the summer. Or in the summer the hay sheds can be vacated and used periodically for meetings and farm tour congregation areas.

This means all structures need to be built with multispeciation in mind. Simple pole structures allow us to quickly change partition configurations, staple up some bird netting, or drive in some additional T-posts to hold hog panels. The whole farm is choreographed for this intricate ballet, in which performers are cued onto the stage at the appropriate time to play their part, and having dramatized their act, exit to await the next scene. Here at Polyface, every day is a theatrical performance. Every day the animals are in different places. And every day the pathogens are shaking their heads, waking up to a whole new stage set, a whole new group of actors.

The result is the pathogens are always lost; they are always confused and disoriented. They are spending so much time trying to survive they don't have time to eat, reproduce, or settle down comfortably. Not on this estate, anyhow. What fun.

In the real microbial world in the Polyface Raken (Rabbit-Chicken) house, Matilda and Harry hatch out in a chicken dropping. She thinks he's a hunk and he thinks she's hot. They decide to get together, marry, build a house, and make babies. They peer out to the horizon, about a quarter of an inch away, and spy a beautiful hilltop to build a home and live happily ever after. They hike over there, spending a couple of hours (10 years of equivalent human life) to get there, only to discover to their utter dismay that it's a rabbit turd. They look at each other in dismay, realizing that they don't have enough time to find a new location and make babies. So they die childless. And that's the pathogen microbial world in synergistic multi-speciated housing situations.

I said nature's sanitation template involves two big protocols. We've just discussed the first one: sunshine and rest. Now let's look at that second one: vibrant decomposition.

Nature is not about sterility; it's about balance. Even Jesus' admonition in the parables to leave the weeds among the wheat lest in pulling all the weeds you pull up some of the wheat, could have application here. If you nuke all the bad guys, you also nuke the good guys. Again, remember this microbial, invisible-to-the-naked-eye world that swirls around us. Our goal should be to create a habitat that allows a balanced community to exist: that allows the good guys to proliferate faster than the bad guys.

The only place we want sterility is in surgery. Otherwise, we simply want a battlefield that tilts the balance toward the good guys. Interestingly, a paper written in 2006 by Callaway, Harvey, and Nisbet at the Food and Feed Safety Research Unit, USDA/ARS (Agriculture Research Service), College Station, Texas put forth the "hygiene hypothesis." The basic idea is that if food is too clean, immunological function becomes lethargic. Then when a real threat comes along, the immune system kicks into overdrive, creating auto-immune disorders like asthma.

Here is a short quotation from the paper: "The hygiene hypothesis was first publicized in the early- to mid-1990s, and has slowly gained currency among MDs, researchers, and public health officials. This hypothesis states that a lack of exposure of children (as well as adults) to dirt, commensal bacteria, and 'minor' pathogenic insults results in an immune system that does not function normally."

The ability to rot is a benchmark of good food. Food that won't rot probably won't feed our internal community. Rotting is decomposition. Anyone familiar with a compost pile knows that you can't build one as small as a 2 foot cube. It needs to be at least a 3 foot cube. This is because the mass has to be big enough to support a core community of critters. When edges are too close to the core community, microbes can't feel safe enough to proliferate. I realize this is an extremely unscientific way to describe the process, but I think putting it in personal terms is often a better way to grasp the big concept. And all of us, at the end of the day, really think with our hearts, not our heads.

In the field, this vibrant decomposition occurs when solar energy produces biomass. We discussed this cycle at length in the earlier chapters of this book. In a housing situation, however, we can duplicate this process with the compost pile principle. The mass issue is why I'm such a proponent of deep bedding. It doesn't matter whether it's a horse stall, goat pen, chicken house, or cattle barn, when animals are being housed, they need deep bedding.

Generally bedding refers to carbon (that's the sponge— sawdust, wood chips, straw, peatmoss), nitrogen (contained in the manure and urine), moisture (generally supplied by urine), air and microbes (usually supplied by the GI tract of the animal). The deeper the bedding, the more functional it becomes. If it's less than about 8 inches deep, it's too thin to support active decomposition. And herein lies the rub: most animal housing structures are not built to handle bedding deeper than about 8 inches.

The result is that every time the bedding gets just deep enough to begin supporting a healthy microbial community, we have to clean it out and start over and that shuts down any decomposition. The components must be in balance. Obviously the bedding can't be too wet or it putrifies like a swamp. It can't be packed too tight or air can't penetrate. It must have a carbon:nitrogen (C:N) ratio of 25-35:1. The ideal is 30:1. Different sources of carbon and different sources of nitrogen have different intrinsic ratios. For example, poultry manure is about 7:1, whereas horse manure is about 25:1. Sawdust is about 500:1 whereas deciduous leaves are about 40:1.

Generaly, if you smell ammonia, you need more carbon. What you're smelling is the vaporization of nitrogen. The bedding should smell like forest soil or leaf mold. Here at Polyface we enjoy telling people that you can have a picnic in our livestock housing—it smells that good and it's that clean. But clean is not scraped concrete. Clean is not sterility. Clean is a healthy microbial community. So all of our livestock housing structures are built to accommodate at least 24 inch bedding, and in most cases, up to 48-inch bedding. Again, the deeper it is, the more vibrant the decomposition. When the cows start rubbing their backs on the barn rafters, we hope spring is not too far away.

Designing farm structures like this is certainly not part of mainline agricultural engineering. Remember the earlier chapters: the whole paradigm in modern industrial agriculture is to pour concrete, scrape, and build lagoons. Deep bedding creates a softness and warmth in livestock housing facilities. Concrete is cold and hard.

The one arena where this is beginning to change is in the swine industry. The Danish system that houses hogs on deep bedding, often under hoophouses, uses a pony wall to hold in the bedding. This is a great development and I encourage its duplication. The carbon needed creates a market demand, and therefore a price, for carbon like sawdust, wood chips, and straw. If all the money spent on fertilizers to compensate for improper carbon disposal had been spent on carbon in the first place, we wouldn't have a carbon sequestration problem. All of these carbon waste streams would instead be profitable business enterprises. A free market carbon trading economy. Now wouldn't that be something?

Deep bedding is good for hooves, claws, and toes. It stays warm in the winter. But most of all, it creates a habitat for pathogen-fighting nematodes to proliferate. And that is the real advantage of deep bedding. Everything else: carbon cycles, real time fertility, solar biomass, manure retention—is all cream.

While everyone else is out wringing their hands and living in fear about the next pathogen attack, we're just dumping in carbon and watching happy animals. And moving animals. And choreographing this wonderful ballet. That, friends, is the sheer ecstasy of being a lunatic farmer.

TAKEAWAY POINTS

1. Michael Beauchamp should be more famous than Louis Pasteur.

2. Nature sanitizes using rest and sunlight.

3. Nature sanitizes using vibrant decomposition.

4. Livestock bedding should be more than 8 inches deep.

5. Sickness and pathogenicity is all about the terrain.

Chapter 13

Sensually Romantic

Who wants to live next to a Tyson chicken factory? How do you know you're three miles from a beef cattle feedlot? You can smell it. And it's obnoxious.

God gave us our senses for a reason. How can you tell if a wound is infected? You can smell it. How can you tell milk is going bad? You can taste it. If it really goes bad, you can smell it. Finally, when it curdles, you can see it. The point is that our senses are our first line of defense in wellness.

To be sure, senses are not an exact science and we have to appreciate differences. I admit to being a wimp when it comes to cheese. I've had the pleasure of eating many artisanal cheeses in my life. Some I really liked and others didn't tickle my taste buds. I haven't found a beer that I liked. Or wine. I drink it politely, but I confess that it all tastes the same to me. Whenever I drink wine, I just tell myself: "time for another worming. Enjoy." It all tastes like cough syrup to me. When wine connoisseurs are talking about the taste and palate nuances, I'm thinking: "so he's describing the difference between Robitussin and Nyquil." But I know wine is good for me, so I drink it. Everybody needs a good worming occasionally.

Several years ago I went with Eliot Coleman to Laverstoke Farm outside London for a three-day think tank huddle. The grand master of Laverstoke, Jody Schechter, convened the gathering, moderated by SLOW MONEY guru Woody Tasch, to vision where our movement would and should be in 50 years. My book *EVERYTHING I WANT TO DO IS ILLEGAL* was a direct result of that gathering.

At any rate, Jody's in-house chef prepared fabulous dinners for us and it was a great time. One evening, however, Jody flew Dan Barbar, internationally acclaimed chef from Blue Hill Restaurant in New York, over to do an evening repast. As I remember the incident, Laverstoke had been dry aging a lamb in preparation for this special occasion. In the British tradition, this was something like 30 days of aging.

When Dan arrived, he smelled the lamb and said it was spoiled. The British chef reared up on his hind legs and a veritable chef cat fight ensued. Dan had insulted the queen. He had insulted the British. Didn't he know this was British meat at its finest? Anyway, the British chef cooked the lamb and they found a piece of pork for Dan to cook. That evening the embarrassing day's events gradually filtered out to the great angst of our gracious and always hospital host, Jody.

Of course, with such an auspicious group, from California's Amigo Bob to New England's Hans Klaas-Martens to Woody to Eliot—my goodness, none of the drama made it past those guys. They are the elite of the elite. I definitely felt outclassed. And so with great gaiety these august cosmopolitan guests gathered 'round the Laverstoke dining table and enjoyed both lamb and pork, as well as everything else. Nobody got sick. We all enjoyed everything. I confess that I liked the pork better than the lamb. It wasn't the last time I've had that British long-aged lamb. I had some at Salone del Gusto in Turin, Italy more recently, and I thought it tasted a bit rotten. Tender, though. Really tender.

By contrast, the lamb I had in Australia, now that's a different story. I've never liked lamb as much as other meats, primarily I believe because in county fairs around Virginia they always serve mutton. And it's grain-fattened mutton. To me, it's strong. But Hispanics love it. When I spoke at some seminars in Australia, the first day I had lamb at a pub and it was beyond wonderful. The whole family was on that trip with me and we all agreed that it was outstanding. The rest of our stay, everywhere, we ordered lamb.

The point is, I understand that a subjective element certainly exists when we talk about taste and beauty. "Beauty is in the eye of the beholder," the old saying goes. But some things can't be made beautiful or pleasurable. Raw manure stinks. Sick animals are not beautiful, no matter how you slice it. Spousal abuse is ugly, no matter how you describe it.

With the introductory caveat that senses are somewhat subjective, I'm going to plunge ahead with the basic concept that food, from production to palate, should be aesthetically, aromatically and sensually romantic. Stench, ugliness, and obnoxious fecal particulate clouds should not assault our senses when we visit a farm. And yet that is the rule, not the exception.

In fact, that is why many farmette developments preclude by deed covenant things like "farming and other noxious land uses." Specifically prohibited are pigs and chickens because most people associate them with the most egregious sensibility violations.

In modern American culture, farms are considered liabilities rather than assets. Nobody wants to live next to one. The whole "Right-to-Farm" movement developed due to the nuisance suits filed against obnoxious farmers. And in every agrarian neighborhood, the farmers defend their odoriferous contribution with great sanctimony. Here are some of the most common arguments:

1. We were here first.
2. You'd better get used to fresh country air.
3. Smells like money to me.
4. Let folks starve, then they'll appreciate what we do.
5. I'm not hurting anybody.
6. Stupid city people.
7. Don't they know this is what it takes to feed people?

In typical circle-the-wagons mentality, farmers everywhere are erecting "No Trespassing" signs. Isn't it fascinating that at the very time when people are yearning for local food and local farm connections, the industrial system is becoming more opaque and more paranoid? Farmers in California who grow salad mix are having to sign affidavits that nobody under 5 years old may visit their farm, in order to keep out potentially pathogenic diapers. Doesn't it say something about our food system when children are not allowed to encounter it? More about that in a future chapter.

The truth is that while farms were industrializing, they were also becoming more repugnant to people's senses. As a result, the locus of production became more sequestered in highly rural areas, out of sight, out of mind, and hopefully out of smell. The environmental and neighbor abuses occurring in these remote areas, where huge production factories have located, are enough to make you nauseous.

Whenever a culture encourages an economic sector to isolate itself from transparency, that sector begins making economic, ecological, and social shortcuts. In direct contradiction to that, I believe production and processing should be so enjoyable and beautiful that kindergarten classes want to come and be a part of it.

And while the industry hides behind the notion that the "No Trespassing" signs are for biosecurity, the truth is that the farmers are really saying: "since our animals and plants have no immunities because our production system has created immuno-dysfunction, we can't let you come here."

Immunities have been compromised due to stress. My own unscientific, experience-oriented formula for stress goes like this: mass X density X time = stress. Think about it. If you have a houseful of people for a Christmas party, that's not stressful. But if somebody came along and locked the door, requiring you all to stay in that house for a month, that would create stress. Clearly, time is a factor.

If, in the aforementioned scenario, only three people were in the house, you probably wouldn't find a month living in confinement stressful. What makes it stressful is the density. Clearly, density is a factor.

But add mass, and everything gets more stressful. A houseful of people at a Christmas party is one thing. But put people at that same density, like people per square foot, in a football stadium, and now you have situations where people get trampled to death. Mass just ratchets up the time and density factors. Now apply these principles to CAFOs. These animals are in these conditions their whole lives, at high densities, at unprecedented mass. The sheer numbers are unbelievable.

In our brooder house here at Polyface, we have it sectioned off so that even if we start a 3,000 bird batch, they are sectioned off in three 1,000-bird compartments. That means when they become frightened and skitter to a corner, only 1,000 can crowd up rather than 3,000. And in big industrial houses, 15,000 is common. They can easily frighten into a group and suffocate the ones on the bottom.

I read a poultry management book that said if in the big industrial houses all the birds were sectioned off in 1,000 bird compartments, half the sickness problems in those factories could be eliminated. But mention such a plan to someone in the industry, and all the barriers go up:

1. That's not efficient.
2. That's a logistical nightmare.
3. That costs too much.
4. That would make it too hard to clean the house.
5. That would be unsightly.
6. That would make it too hard to walk through the house.

I daresay that if the industry could bring itself, just once, to think about what would make the chickens healthier and happier without drugs, they might discover that partitions would pay for themselves.

Animals are a lot like people. They want fresh air, exercise, and sunshine. As minimal and basic as those requirements are, they are denied to 99 percent of the animals in America. To be sure, 1940s style mud lots and dirt chicken yards are not the answer. Fresh air, exercise, and sunshine without clean ground won't do a bit of good.

As farmers, though, our responsibility is to provide those basics on clean ground. And even when they have to be housed, temporarily, for inclement weather or extra care, our structures should be roomy, airy, and sunny. That's one reason here at Polyface that we have our winter laying hens in hoophouses. It's a wonderful place to work, just like a greenhouse. Few buildings are as fun to inhabit in the winter as greenhouses. The ambiance and gentle warm solar-heated air are magnetically inviting.

Good farming should be like that. The whole ambiance should attract us to it, rather than repel us. Children should be attracted to it, not repelled. Mark it down, if it smells bad or it's not beautiful, it's not good farming. What's good for the senses is good for the animals and good for the plants. Indeed, it's good for the ecosystem.

If this one rule were applied to modern food production in America—aesthetically and aromatically pleasant—it would fundamentally change the entire agricultural paradigm. When I drive along a country road and smell insecticides or herbicides, I try to hold my breath for as long as I can and hope I drive out of the area.

As we wind this discussion down, I have to address the axiom that when I say beauty, for the most part I'm talking about biological beauty. While I very much enjoy and appreciate well-built structures and ordered infrastructure, that is second to the glow and radiance of the fields and livestock. Actually, profitable farms exhibit a somewhat threadbare appearance. At our farm, we don't worship white board fences and immaculate outbuildings.

Polyface will probably always have a bit of an in process mystique and appearance. We are constantly changing, adding, refining, innovating. An environment conducive to innovation exudes a somewhat disoriented look. Louis Bromfield, icon of early sustainable agriculture and man of great wealth, said he liked his mansion-type farmhouse to look like it had been added on and remodeled, like it hadn't arrived. Innovative farms will never have that pristine, perfectly constructed look like a farmstead built with off-farm wealth. Buildings don't make a farm. Fencelines don't make a farm. Spending doesn't make a farm.

The vibrancy and vitality evident in the plants and animals are what create the aesthetic and aromatic appeal. When visitors come to the farm, I don't show off the buildings. It's a functional farmstead. A working, profitable place. We let fencerows grow up in wildness to provide habitat for critters. Most farmers around here spray their fencerows to create a pristine, ordered appearance. I don't think that's pretty, because you'll never see a cardinal or blue jay chirping away in the tangle of some vines or bushes in the fence. Brushy fencelines attract wildlife.

To farmers obsessed with dominating the landscape, the wild areas around ponds and field edges indicate negligence, lack of discipline, and weed seeds wafting across the neighborhood. But to me they show a respect and appreciation for nature's order, and ultimately express a value in maintaining at least some areas where I withdraw my hand. Sometimes those are the most interesting spots to visit on the farm. And no, I don't feel compelled to subjugate them all to my dominating hand. A little bit of wildness is good for the soul.

Over the years, as we've fenced out steep hillsides, our once open-vista farm has taken on a more patchwork appearance. We've created wild zones like fingers, out into the open areas, to stimulate wildlife runs and wildlife penetration into the open land. Some of these runs follow steep slopes and others follow riparian areas. But the patchwork now creates its own beauty. Behind and around each wild area is another field, another eggmobile, or herd of cows. This makes a surprising, almost explosively spontaneous and discoverable farmscape that both delights and enthralls visitors. And makes a wonderfully pleasing place to work.

Think how different a Tyson chicken factory is. Same old same old. Day after day after day. Fecal particulate. Unhappy chickens. Stench. A real downer. When I walk out to the chickens, however, I might encounter a fawn nursing its mother, or a wild duck with fresh hatchlings in tow gliding across the pond. In the early morning, with the sun shining through a thousand dewy diamonds clinging furtively to crisp orchargrass and white clover, I can't imagine a more beautiful environment. The dew breaks the sun rays into mini-rainbows that kaleidoscope across the pasture. It's enough to stop your heart, and make you want to tiptoe across the field to leave such splendor undisturbed.

That's the aesthetic and aromatic sensual romance I'm talking about, that draws you back and draws you back, not because you'll go bankrupt if you don't keep coming back, but because you just love to be there. Communing. Relishing. Anybody who feels and senses that romantic intoxication with a Tyson factory chicken house is just weird. I'm sorry. I know beauty is in the eye of the beholder, but that only extends so far. If the farmer has to argue you into believing it's pretty, something is wrong. It ought to appeal on its own merits, without regard to argument.

We've had people who thought Polyface was horrid. Some animal rightest radicals came by and said it was abusive to put our chicks in field shelters. They said the chicks should free range. Apparently they haven't seen what crows enjoy doing to little chicks in the field. Or 'possums. Or cold rain. It's not a pretty sight.

And certainly some people have written completely false information about us. You certainly can't believe everything you read. I realize that by claiming we're beautiful that puts us in a precarious position with people who enjoy shooting others down. All I can say is "Come." If you don't believe it, come. And I encourage people to visit their farmers. Whoever supplies you with food, visit them. You'll get a sense pretty fast if it's for real. Trust your senses.

The steady diet of sensual beauty that I take in every day certainly gives me the sheer ecstasy of being a lunatic farmer.

TAKEAWAY POINTS

1. Trust your senses at least as much as the science.

2. Farms should be aesthetically and aromatically attractive.

3. Wildness patchworks add interest and discovery to the landscape.

Chapter 14

Less Machinery

Farmers love machinery. You don't have to go to very many county fairs or farm shows to realize how much farmers love machinery. It's practically an obsession.

The rural joke is you can make fun of a farmer's wife, but just don't make fun of his tractor. A friend of ours had a sign hanging in the kitchen: Wanted: Farm Wife. Must be able to cook, sew, clean, butcher. Must have tractor. Please send picture of tractor." Isn't that hilarious? But it about sums up how we farmers view our equipment. Fortunately for us, the Ten Commandments prohibit coveting "your neighbor's wife," but not his tractor. We'd be in hot water if God had included tractors.

You know what happens when you play a country music song backwards? You get your truck back, you get your tractor back, you get your dog back... The old saying that the only difference between men and boys is the price of their toys certainly holds true for farmers. Why do you think John Deere is green? It's the alternative investment in rural America.

I feel like Charles Dickens' beginning *A CHRISTMAS CAROL* with the line: "Marley was dead to begin with." Farmers

172

love machinery to begin with. No matter what happens with the animals, the crops, the soil, if a farmer can get on his tractor and feel the power shudder through his thighs, all is okay with the world. After all, most farmers become farmers in order to run machinery. Look at little boys—and sometimes little girls—on equipment at the county fair. Forget the rides. Just put them on a big tractor and they're good for a half hour.

This is why I think earth friendly farmers have to be sissies. I'm a sissy farmer because I don't worship machinery. After all, what man wants to come in from working out on the farm all day and when his lovely Matilda asks: "Harry, my big hunk of a farmer, what did you do all day?" must respond: "Oh, I made the cows happy." That's no way to be. A man would respond, in dissonant gravely staccato: "Oh, I tore up the back forty, killed 40 million earthworms—plowed right through 'em—put on ten gallons of Bicep herbicide and spent all day atop pig iron under my thighs. Here, smell the diesel fuel." Then pounding his chest, he says: "I'm a man!" And Matilda agrees.

It's just not manly to be more concerned about happy chickens than bigger chickens. It's just not manly to be more interested in pigaerators than cultivators. It's just not manly to be more interested in earthworm microbes than lagoon irrigator guns. After all, we've got masculine dignity to maintain here. We can't be going around talking about pigness and cowness and tomatoness. What kind of feminine sweet talk is that? No self-respecting man uses that kind of lingo. Men are supposed to control, dominate, pontificate, and resonate. What's this talk about caring, nurturing, and massaging?

Realize, then, that when I say at Polyface we think the less machinery we have the better life is, that's loony in the first degree. I think machinery is the option of last resort. After you exhaust all the alternatives, then you can buy equipment. Much, if not most, of the machinery farmers have is unnecessary. Machinery you don't own doesn't have to be maintained. Machinery you don't own will never break. Machinery you don't own never has to be replaced. Machinery you don't own lasts forever. My favorite machinery is the machinery I don't own.

When you're as lunatic fringe as we are at Polyface, half the machinery current manufacturers make doesn't apply to anything we do. We've had to make many of our machines. Some have worked and some haven't. Growing up, when we got ready to try something new and we were discussing all the unknowns, Dad would liven things up with his incisive prophecy: "I don't know about this, but we're going to know a lot more in 30 minutes."

Half of our machinery is obsolete when we buy it. When machinery is obsolete, the price goes down to about scrap metal value. So all you have to do to get really good machinery deals is to stay a generation behind the times. Then you get all the castoffs of the progressive industrial farms.

I remember an auction Dad and I attended. We needed a new hay mower. At the time, the new rage was a machine called a mower-conditioner. It combined mowing and conditioning (crimping the fresh-cut grass between two rollers to encourage evaporation of plant juices) in one machine rather than two machines. The accelerated evaporation from crimping helps the forage dry faster. I never had much use for crimping because if the weather is good, it dries anyway. If the weather is bad and you get some rain on it, the crimped loses more of its nutritional value because the rain washes over the crushed stems.

Since mower-conditioners were the rage, simple sickle bar hay mowers were obsolete. Dad and I saw one advertised at a nearby auction so we went to bid on it. The auctioneer started at $500, which we would have been happy to pay. A new mower would have cost us $1,500. But Dad waited, not wanting to be the first bidder. The auctioneer came down to $400, then $300, then $200. Dad started getting antsy, afraid that if it went too low everyone would jump in thinking it was too good a deal to pass up.

Auctions offer real psychological educations. When the auctioneer dropped it to $100, Dad couldn't stand the pressure, so he jumped in and bid. Try as he might, the auctioneer could not get another single bid. "Sold, for $100 to Salatin," he said, and went on to the next item. Dad looked at me sheepishly and said: "If I'd waited, we might have gotten it for $25." Ever since then, I don't think I've been the first bidder on anything. But it shows the kind of deals you can make if you remain content with everyone else's castoffs.

We brought that mower home, greased it up, and used it for many years until it finally fell apart. It worked great. I can only imagine the discussions around our neighbors' dining room tables that night about that crazy Salatin guy buying that old mower. I'm sure they shook their heads in bewilderment: "Why would anyone want something like that? Something that old. Why, that machine is obsolete. No self-respecting farmer would own it, much less buy it."

In the early 1960s, when all farmers had switched over to hay balers, Dad was going the other direction. Back to the future. This whole chapter in my life probably occurred because of Dad's frustrations with an old Case baler that was here when we bought the farm. It was a constant frustration. It must have been built in the 1950s sometime. John Deere and New Holland definitely had the better balers at the time, but this one came with the farm so it was virtually free.

For you uninitiated readers, hay making has three parts. First you mow tall forage (grass, legumes, or combinations of the two). After it dries, you rake it into a windrow. Then you bale it. That puts it into a package you can move around, stack, and handle. At that time, bales were little square bales about 3 feet long and 1 1/2 feet wide and 1 1/2 feet high. If the hay was in good dry condition, the bale would weigh 40 or 50 pounds and was tied with two strings. The automatic knotter was the thing that made balers really become popular. Initially, two people had to ride on the back of the baler and thread wire through a hole and twist it, then cut it. The wire held the bale together.

But eventually some sharp engineers invented two needles, just like slightly curved sewing needles, about 2 1/2 feet long that would carry string (twine) up into the knotting mechanism when the plunger (the heavy metallic fist that packs the hay in a square tube, called a chamber) was on its forward stroke. All of this mechanism would tie a knot and the needles would retract before the plunger returned to compress another wad of hay. A metal hand on a cam feeds the wads of hay into the chamber. The plunger has a knife that sheers off the wad of hay as it's compressed against the packed wads already in the chamber. Each wad is called a flake.

Not only was the knotter a very sophisticated set of gears, bill hooks, and knives to cut the twine, but the timing of the needles had to perfectly mesh with the forward stroke of the plunger. Otherwise, the plunger could break off a needle. That not only put you out of business for the day, but also cost a lot of money. Each needle was worth two days' wages for an average working man back in 1962. I don't know if our baler was a lemon or if it was just the nature of the beast at the level of engineering sophistication available in that day, but I know it missed as many bales as it tied.

Because of the baler's inconsistent knot-tying performance, my older brother Art or I would ride on the back of the baler and watch the knotter. If it tied, we'd give a thumbs up hand signal to Dad, who always drove the tractor. If it didn't tie, we'd cross our arms in a big X and he would stop, come back, and tie the knot by hand. After Dad passed away, I was going through some of his files and found his amazing drawings of the knotters, with pages of handwritten explanations as to how they functioned.

I can't imagine the hours of frustration and consternation he must have gone through during those first couple of years, trying to understand this machine, and trying to get all the settings correct so it would work. Obviously this frustration set him on a course to find an alternative. Balers like this had only been widely adopted for about ten years at that time. When he was a boy, farmers forked hay on a wagon with pitchforks. He would tell stories about the neighbor men who would come over and fork the hay onto the wagons. They would put enough loose hay on the fork to bend the pitchfork handle when they tossed the hay up onto the wagon.

Forking hay is truly an art, just like all work. I remember noticing in my late teen years that I, too, could bend a pitchfork handle with the load of hay I had on it. That's one of those rites of passage into manhood that you remember when you get as old as I am. My brother-in-law, super athlete growing up, wears a T-shirt that says: "The older I get the better I was." Between the forking up years and the hay baler, from roughly 1925 until 1953, a machine was invented and widely used called the hay loader.

This machine was essentially an inclined plane. Towed behind a wagon, it straddled the windrow, picking up the hay with a rotating pickup very similar to modern balers, and pushed the hay onto an inclined plane of sheet metal. Several boards with swiveling tines on the underside rotated on a big cam and pushed the hay up this 10 foot inclined plane, where it dropped off onto the back of the wagon. With a pitchfork, a worker would shepherd this loose hay onto the wagon, spread it out, tromp it in, and gradually load the wagon. This made the picturesque loose hay wagons that look like a big rounded bread loaf, with the hay edges hanging down to the ground.

A neighbor had one in mint condition in his barn and Dad bought it for almost nothing. It saved the neighbor from having to take it to the dump. From the mid-1960s until probably 1980, we made loose hay. I have to be the youngest farmer in America who actually knows how to put loose hay on a wagon from the mouth of a hay loader. Another neighbor had a hay fork, and since he had long since quit using it, he just gave it to Dad. A hay fork is a four-pronged contraption with a trigger on it that you trip by pulling a rope.

In barns built for this, the hay fork would be plunged into the loose hay and the fork would ride on a track hanging in the peak of the barn. When the load of hay got to where the operator wanted to drop it, he would yank the trip rope and the hay would fall into the mow. A team of mules, horses, or oxen, and then later a tractor, would pull this hay fork along the track with a rope that extended out the other end of the barn. The animals would get used to the rope going slack when the hay fell off, and automatically stop and back up so the operator could pull the fork and carriage back to the wagon load of hay and take off another jag of hay. About ten forkfuls would clean off a wagon.

In our pole barn, we didn't have a track. Dad affixed a pulley to the final rafter up in the peak of the barn. By stationing a signaler in the barn window, the fork operator could ready the fork and yell "Up!" The signaler would give the up signal. When the operator pulled the trip rope, he would yell "Stop!". At that point, the signaler would extend her arm straight out. When the fork operator had the fork clear of the hay and ready to pull back, he would yell "Down!" The signaler would bend her arm down, and the person driving the pickup or tractor would come back slowly while the fork operator pulled the collapsed fork back to the wagon for another load.

This was the way we made hay all during my teen years. As the 1960s gave way to the 1970s, farmers began buying round balers. That was the new rage. One of my fondest memories of Dad, and the kind of story that illustrates his zest for unconventionality, was the day we were plodding along the meadow making hay. He was driving the ancient Oliver 88 tractor and I was stacking the wagon. Along came one of our neighbors in his snazzy spanking new Vermeer round baler.

Dad stood up on the tractor, turned around with a great big full-on smile, and yelled over the tractor engine, pointing first at the neighbor: "Old!" Then pointing to me: "New!" We were using a 50-year old machine. Archaic. Obsolete. But you know what? We could fix anything on it. If anything broke, we could fabricate the broken part in the shop. It was simple. And it wasn't worth a penny. No taxes. We were making hay with what everyone else considered a pile of scrap metal.

And it was good hay. We carried a bucket of salt on the endgate of the wagon, and every so often I'd scoop a little and flick it on the hay. The combination of salt and looseness (aeration) made for perfectly cured hay. If we ever could have figured out how to efficiently unload it, I'm not so sure we wouldn't still be using the hayloader today.

As our fertility increased, the herd size increased. Haymaking increased and we could no longer get the wagons unloaded in the barn. We didn't have a barn conducive to loose hay so we couldn't put it in a cube like it is in mows made for loose hay storage. Our setup was an inclined plane. Bales stack nice and square. We eventually had to convert to square bales in order to use our roof space more efficiently. And bales did speed up the haymaking process. But I remember those days with great fondness, and that machine sure made good hay.

Today, with our pastured pigs, we use Grain-O-Vator buggies that were used by thousands of farmers 40 years ago to feed range turkeys and outdoor pigs. But as both pigs and turkeys went into buildings with automatic feeding systems, these buggies fell into disuse. We've picked up several at virtual scrap metal prices. They are perfect for our pastured poultry and hog systems.

Isn't it amazing that the backbone of our machinery fleet was state-of-the-art half a century ago?

We try to squeeze every use out of every axle and chassis on the farm. When Dad junked that old baler, he saved the axle. We had a dump bed that was on a 1951 International dump truck he bought shortly after we came to the farm. He pulled that dump bed off and put on a diamond steel bed so he could haul pulpwood. When that baler axle became available, we mounted the old dump bed on it and had a nice steel trailer. We welded the old baler hitch on the front of the dump box and it made a fine utility trailer to use around the farm.

We bought a grain buggy from a fellow, but instead of being a trailer, it was on a wagon chassis, which made it very hard to back. We had a defunct manure spreader we'd recently decommissioned, so we separated the manure spreader box from its axle, took the grain buggy off its chassis, and mounted the grain box on the old manure spreader axle. It doesn't look like much, but it's now a trailer and you can back it up easily.

That whole project left us with the old heavy-duty wagon-type grain buggy chassis. We needed a forwarding cart to bring logs out of the mountain. We'd been using hay wagons, but they were cumbersome and not heavy duty enough. In addition, since they were just a flat wagon with no sides, too often a log would roll off the side before I could get it chained down. I took that grain buggy wagon chassis to the shop, welded four heavy pipes on an incline, like a big cradle, and we've used it for years as a log forwarding cart. Heavy chassis, maneuverable, extremely functional.

The average farm requires about $4 worth of buildings and equipment to generate $1 in annual gross sales. In other words, a farm generating $300,000 in annual gross sales, on average, is operating with roughly $1.2 million worth of buildings and machinery. Do you know what our ratio is? It's 50 cents to $1. For you percentage-challenged readers, that's an 800 percent difference. Think about that for a minute. Isn't that a remarkable difference?

I introduced pigaerating earlier, mainly due to its respecting the pigness of the pig. And it makes great compost. But perhaps the greatest benefit is economic. When animals replace machines to do the work, the farm's profit potential becomes size neutral. The old axiom that "you have to get big or get out" applies when you're moving material and using machinery. The fact is that a 3-cubic-yard front end loader is much cheaper to operate per pound of material handled than a 1-cubic-yard front end loader. The big cost is the operator's labor. If the operator can move a lever and pick up 1,000 pounds rather than only 500 pounds, the whole procedure becomes much more efficient.

The difference in cost between a 40 horsepower tractor and a 60 horsepower tractor is nominal. Those extra horsepower don't cost that much more than the initial ones. Both machines have pistons, engine blocks, radiators, alternators, starter motors, and transmissions. The machining and cost differential between making those parts for a smaller tractor compared to a larger one is negligible.

When you're in the materials handling business, it is cheaper by the ton and cheaper by the million. But if animals do the work, you don't have to turn the volume to recapitalize the infrastructure that rots, rusts, or depreciates. If animals do the work, the profit potential becomes size neutral because you're using appreciating equipment. That revolutionizes the profitability of a small farm.

Instead of double handling all that livestock bedding and then turning it with a $20,000 sophisticated single-use machine, using my time, and using diesel to power it, the pigs do all this work on their own time. I don't have to steer them. They do all that work and I don't have to drive them anywhere. I can just come in and read books while the animals do the work. This model works just as profitably per ton of material whether I use two pigs or a hundred pigs. Size makes no difference.

To me, that's one of the benchmarks of breakthrough business models. When a protocol does not depend on scalability for profitability, you know you're onto something revolutionary. And how much more fun is it to do all that work with hogs, to see them cavorting around, to scratch their ears and rub their bellies when they're napping, than it would be to climb onto an iron monster and chug it down through a compost pile. No comparison, my friend. Sheer ecstasy.

Sometimes on our farm we like to have fun with the neighbors, just to confirm their opinions of us as lunatics. Kind of like renewing your wedding vows. It does good once in awhile to do something clear off the wall just to keep the neighbors talking and shaking their heads. Kind of play with their minds.

We don't own a seed planter. The only planter we have is a little hand-cranked cyclone seeder made to sling over your shoulder on a strap. You dump the seed in a little bag atop this contraption, hold it against your stomach, and turn the crank. A little paddle flings the seed out about 15 feet, depending on how heavy the seed is. That's the only seeder we have. We use it to plant pasture in forest-to-pasture conversion areas.

In half a century, on this farm, we've converted about 60 acres of open land to forest and about 10 acres from forest to open land. We don't bulldoze out the stumps. Instead, we cut the stumps as low as we can, stack all the branches in a pile and either chip or burn them, then broadcast seed out over the ground with this cyclone seeder. In order to get good germination—what's commonly called a good take—the soil surface needs to be disturbed enough to get it and the seed in good contact. Ideally, the soil actually covers the seed about a quarter inch deep.

But how do you do this on rough ground with stumps scattered all over? For years, I've answered this need by just cutting a couple of branches and dragging them over the seed. I walk over the area with the cyclone seeder, tie some branches to my waist with some rope, and do both operations at once. If it's a bigger area, we use the 4-wheeler (All Terrain Vehicle—ATV). We tie some branches behind the 4-wheeler and then one guy rides facing backwards on the seat and cranking the cyclone seeder. Everything in one pass.

On bigger areas we've used a tractor. Of course, the tractor can pull a bigger group of branches, or even a couple of treetops, which disturbs better and insures a better take. Several years ago we leased a farm on which the previous operator had planted and harvested a 10-acre corn field. When we arrived in the spring, that 10 acres was just dirt and cornstalks. Guess how we planted it? That's right, our little cyclone seeder and tree tops.

We grazed the mob there the following winter and the fields were covered with cow patties that we wanted to spread out.

We hadn't gotten an eggmobile there yet. What to do? Most farmers buy pasture drags, kind of big metal mats, for this purpose. Resourceful farmers hook several tractor tires together and pull them around to bust up the cow patties.

One area, about an acre, right down by the main two-lane highway, had been pugged up during a winter rainstorm. The mob huddled up to stay warm and in the ensuing rain tromped the area into mud. In our experience, these areas heal up very nicely since the cows are only there for one night. In New Zealand, they call this land treatment "deep massage." But most of our landlords think this temporary muddiness is unsightly, and this one was no different. "Could you get some seed spread on there?" he asked.

"Sure." We wanted to spread cow pats on some of the other fields anyway, so we went over with the pickup and trusty cyclone seeder. I cut a couple of medium-sized trees and we chained them to the back of the pickup. We waited until the top quarter inch of soil was dry. Not only did we not want to track up the field with the truck, but we also didn't want to smear the soil surface. We wanted some dryness so the soil would roll up over the seeds.

We dropped the tailgate on the pickup, set a guy on the spare tire cranking the cyclone seeder, and off we went, tree tops crazily rolling and bumping along, stirring up a cloud of dust. We perfected our technique out of sight of the public road and once we had everything working perfectly, rolled over the hilltop and swooped down on that one acre dirt patch. Boy, howdy! Old codgers driving by in their pickup trucks jerked around to see what was happening.

What a crazy outfit. Pickup bouncing through the field, guy sitting on a spare tire in the back, flailing around cranking a cyclone seeder, and treetops careening wildly, throwing up a cloud of dust. I was laughing so hard I could scarcely drive. I still can't believe nobody took out the fence along the edge of the road. Trucks stopped. Farmers hung out the windows. I'll bet we created some business for community chiropractors that day with all the gawking whiplash. When we got done, we were having so much fun watching the neighbors' reactions, we made a couple more passes right up along the highway fence just for fun. After all, we wanted to make sure the neighbors got their stories straight when they told their knee-slapping tale to their families at supper that night.

I wonder who had the most fun: us watching them watching us, or them trying to figure out what we were doing, coming up with some harebrained conclusion, and then passing along the lunacy to others? I think we probably had the biggest laugh. What great fun, to just play with the neighborhood sometimes. Now wouldn't it just take all the fun out of it if we used a conventional seed drill? How boring is that?

If we used a real honest-to-goodness machine, it could break down. Then we could buy spare parts and spend half a day fixing it. And we could spend time buying it. We could whine to the banker about how "thar ain't no money in farmin'. We could put it on our personal property records and pay taxes on it. Oh, the levity we're missing. I can't stand it.

Here's the bottom line: while everybody else is strutting their machinery around, we're trying to figure out how to do things without any machinery. And that's the sheer ecstasy of being a lunatic farmer.

TAKEAWAY POINTS

1. Profitable farms have a threadbare look.

2. Most farmers love machinery, and machinery is costly.

3. Letting animals do the work eliminates scale as a factor of efficiency and profitability.

4. I'm a sissy farmer.

Chapter 15

Nativized Genetics

"What kind of cows do you have?" That's one of our most frequently asked questions. The corollaries are there too: "What kind of chickens do you have?" "What kind of pigs do you have?"

Genetics have been a fascination with people ever since Jacob put sticks in the water trough and created a more productive flock of speckled and spotted sheep and goats out of his uncle Laban's flocks. Interestingly, in that Biblical story, the number one benefit appears to have been reproductive fertility. He was after reproductive efficiency.

That is a far cry from industrial agriculture's goal, which is volume production per animal. The best illustration of this is in the dairy industry, where milk production per cow's lactation is the holy grail. But in nature's accounting system, as the dairy industry pushed the envelope toward production, reproductive capacity dropped. Since birthing is what creates lactation, reproduction is fairly important.

But with reproduction plummeting, industrial dairy cows average scarcely two lactations. That means the cows are burned out after only two calves. And since half the calves are bulls, a two-calf per cow reproductive rate is not even enough to maintain the current number of milk cows. That is why heifer prices have gone through the roof. Industrial dairies depend on less industrial dairies to have more lactations per cow and therefore enough extra heifers to subsidize the industrial dairies. Often farmers who move to a grass-based dairy program make almost as much money selling their excess heifers as they do selling their milk. And their cows often stay fertile for a decade or more, producing a calf every year.

The same production goal, though, permeates every sector of modern American agriculture. What genetics do beef cattle producers desire? Big calves. Big weaning weights. Vegetables are selected for production per plant, or production per acre. Nobody asks what this production per unit selection process does to eating quality, nutrition, and susceptibility to sickness.

After all, eating quality doesn't matter because it's all processed anyway into some boxed or canned conglomeration. Taste and texture don't matter anymore. Nutrition doesn't matter because we can take vitamin supplements to compensate. And sickness doesn't matter because we can compensate for genetic weaknesses with pharmaceuticals.

Ranching for Profit teacher extraordinaire Dave Pratt makes the point over and over again when he does seminars: "Production does not equal profit. ONLY profit equals profit." And he's exactly right. Whenever you chase one genetic trait to the exclusion of everything else, it creates profound deficiencies in other areas. Grain farmers who receive trophies for "Highest Production Per Acre" of corn or soybeans or barley will always admit, off camera: "I did this on one test acre to win the award. I could never afford to do it on the whole farm." Even mechanics know that an engine never runs most efficiently at full throttle. And yet that's exactly what the farm sector demands of plants and animals.

The inputs in fertilizer, irrigation, foliars, and herbicides to manicure that one acre to win the production award could never be profitably duplicated over the rest of the acreage. The same is true for cattle who win shows. The coddling and attention to those animals would never be practical in a real life production sense.

Breeds of cattle have now been developed in which almost every single calving is an assisted birth. That level of intervention can never make economic sense. But the double muscling is freakish enough to create a sensational story for which people willingly pay extra.

Hybrid corn contains some seven fewer enzymes than open pollinated corn. Of what value are those enzymes? I have no clue. Nobody knows. But I have a sneaking suspicion they are important, or God would not have put them there. For the industrial food system to cavalierly dismiss these genetic nuances as unimportant shows not only an arrogant spirit, but also a naïve attitude.

If we were going to pick one trait as the most important in a genetic selection process, I would pick reproductive ability. After all, if something can't reproduce, it doesn't matter how big it is or how fast it can grow. If conception doesn't happen, or if live birth doesn't happen, everything else is useless. When farmers brag about their big birthweight calves, my response is this: "I'd much rather have one live 50 pound calf than a tractor trailer load of dead 90 pounders."

Commercial turkeys can't even naturally mate any more. When you saw job advertisements for milkers, historically, that had to do with dairies. Now you see help wanted ads for turkey milkers. Turkey milkers? Yes, they milk semen from toms and artificially inseminate the hens. How would you like that for a job? Pretty convoluted use of the word milkers, I'd say.

Decades ago on our farm I dabbled with hot shot genetics. First, I tried some artificial insemination, using semen from a well known supplier. We bred some cows and ended up with four heifers out of the bunch. They all looked good as they grew. We wanted to keep all of them for replacement cows. But two would never breed and one lost her calf at birth. Only one actually bred and delivered a live calf. And she only lasted about four years. A cow should have a calf every year for 10 years. In the final analysis, every one of these heifers from the stud farm was a dud. Every one. He might have been a stud, but his daughers were duds. I'd say that makes him a dud stud.

I've concluded that we shouldn't be doing artificial insemination. Here's why. In natural service, one ejaculation produces one calf. In fact, the bull enjoys several, but finally, when everything with the female is just right, one service produces one calf. In artificial insemination, technicians create 100 service straws out of one ejaculation. That means rather than that one strongest sperm being the fortunate winner and impregnating the egg, 99 lesser also-rans also get to impregnate an egg.

I don't see how that can increase genetic viability and health. From a statistical standpoint, it seems to me that over time you would sacrifice viability. I'm not a geneticist and I'm sure some people will think me completely lunatic for taking this position, but when you realize you're making 99 conceptions from second-rate sperm, it should give us pause. Perhaps this is an area where human cleverness has overrun its headlights. Perhaps if we looked at natural service as a safety valve against over-rapid genetic changes, we would have a more balanced progression. Horror of horrors, I wonder if this applies to humans?

Why can't we humbly accept a closed door (womb)? Why must we assume that if we jack it open, shove it down, push through it, then it's noble for us to do so? Why does "no" always seem like an insult? It doesn't have to be. It can, in fact, be a wonderful protection. Learning no; learning limits; learning boundaries is a healthy thing.

But I, like most people, was not willing to accept this. Our next attempt was to buy a hotshot bull from one of Virginia's finest experiment stations. We bought the top gainer in his class and turned him out with the cows. The next two years we ended up assisting about 50 percent of our calf births. He was a fast grower, but half of his progeny, were it not for our intervention, would have never seen the light of day. And many of those births would have killed the mothers in the process had we not assisted the birthing.

As I've progressed in my own thinking, I now believe any cow that needs birthing assistance should be culled. That's a fatal genetic flaw. I don't help deer and squirrels. Why shouldn't I ask of my domestic animals at least as much as nature affords? The more I intervene, the more crutches I'll need to prop up the weaknesses. A cattle enterprise is not about my becoming a slave to their needs. It's about their doing their part and I'll do mine.

With all this discussion about artificial insemination, I have to at least mention that it's not nearly as fun for the animals. When I see those cows standing around in heat, pleading for service, I can't imagine that my getting them into a headgate and pushing a straw through their cervix is as satisfying as having that bull jump on them. And for the bull...well, enough said. How would you like to ejaculate by electric impulse—into a jar? Come on, people, we're talking quality of life here. How about the cow's quality of life? And the bull's? 'Nuff said.

The fact is that every time genetic selection goals narrow to production per animal, like speed of growth or amount of milk, weaknesses show up in other areas. Fertility is first. Then skeleton, including gait and legs. I want balance. I never want the biggest or fastest. I want balance. That's health.

Kit Pharo, who raises grass-based seedstock for beef producers, says that if cattle producers would only select bulls from cows more than eight years old, it would fundamentally change the functionality of U.S. beef cows. Everything in the industry, though, is based on young. Supposedly the genetics from today's young cows are superior to the genetics of cows born a decade ago. I disagree. Kit's point is that if you let longevity drive genetic selection, over time you end up with functional genetics. A cow that has been in the herd for a decade is inherently one that is functional. Functionality means balance.

The mere fact that she is still in the herd shows that she has staying power. She knows her job and performs. By only selecting bulls from older cows, it protects us from our own jaundiced prejudices about what the ideal animal should look like. The ideal animal is the one that functions. Period. A healthy animal adapted to its region is the one we want, regardless of color, size, or whatever.

This was the secret of Tom Lassiter's Beefmaster breed. He created six criteria for genetic selection, and stuck with it. By having a balanced set of selection parameters, it protected him from personal prejudice. That way nature becomes the defining force, just as it is in the wild.

Here at Polyface, we use non-hybrid dual-purpose breeds for our laying flock. Right now, we're using Rhode Island Reds, Barred Plymouth Rocks, and Black Astralorpes. Those are traditional breeds that would have been found on any American homestead a century ago. These don't lay quite as well as the hybrids like Dekalb Golden, Golden Comets, Cherry Eggers, J.J. Warren Cross and Sex Links—all crossed with Leghorns.

But the hybrids have a couple of shortcomings. First, they lay too prolifically for their metabolism to keep up. These smaller-bodied birds, laying six eggs a week, equivocate to a 150 pound person losing 15 pounds a day. Can you imagine trying to keep up with that regimen? This lay rate does a couple of things. First, the bird begins cannibalizing its own skeleton to keep up with egg shell calcium and other egg nutrients. Second, the birds can't ingest enough green material to make the dark, rich yolks characteristic of nutrient-dense eggs.

Their total energy requirements are so high, they can't sacrifice digestive energy to metabolize something as low in energy as grass and clover. They want corn. The non-hybrids like the ones we use only lay five eggs a week and their bodies are half again as large (5 pounds compared to 3 pounds). Their production would equivocate to that 150 pound person losing 5 pounds a day. While that is still high, it's doable. Although I haven't seen scientific studies to back this up, intuitively it seems to me that a bird that's not cannibalizing her body to keep up with production will pack a little more nutritional punch in her eggs. That just seems reasonable.

Another characteristic difference we've noticed is that the bigger, traditional birds are less flighty. That means they stay in the feathernet (electric fencing) better. They don't go as crazy if they are frightened. They are smarter—they actually watch for hawks. They are hardier—they handle hot, cold, rain and wind better. It stands to reason that they wouldn't be as fragile as the smaller, higher strung birds.

Polyface eggs have a dominant reputation in the mid-Atlantic area. I think one reason is that we have not succumbed to the lure of hybrid egg laying genetics. Yes, it means our production per bird is lower by about 25 eggs per hen per year. Yes, it means our birds eat a little more grain per egg. But, it also means our birds live longer, stay healthier, eat more forage as a ratio of total diet intake, and yield a wonderful carcass as a stewing hen. Most pastured egg operations are using the hybrids. Some even use debeaked birds, which are at a distinct disadvantage for eating bugs and forage.

One of the problems we're having at Polyface is that the nationwide breeding stock flocks for these minor breeds is so small that when we order 3,000 pullets from a hatchery, they can't come in one group. They have to be spread out over a month. That's a nightmare because instead of starting all the chicks at once in the brooder, they come in staggered. They grow at different rates. They are ready to go to the field at different times. They begin laying at different times. Instead of all starting at once, they dribble in over a long time.

This is one reason I'm encouraging farmers to use these minor breeds, because that's the only way the breeding flocks will increase to accommodate larger commercial numbers. When a customer buys a Polyface egg, the ripple effect is huge. That egg keeps a small pasture-based farm in business. It puts a minor breed bird in the field—a smarter, minor breed bird, mind you. It creates a marketable stewing hen at the end of productive life. It patronizes an independent hatchery, which is key in preserving non-industrial poultry. It creates market demand for traditional genetics and encourages existing seedstock flocks to stay in business. That's a pretty cool ripple, don't you think?

As an aside, let me address the radicals in the animal rights movement who keep lobbying and trying to criminalize the shipping of poultry as inhumane. Birds are not mammals. Chicks are fine for up to 72 hours without food and water. As soon as they eat or drink, however, they need to eat and drink several times a day. Ideally, 48 hours is the cut-off, but they will hang in there up to 72.

I was speaking at the Washington D.C. Live-Green expo recently and during Q/A a lady asked if I endorsed Murray-McMurray shipping chicks around the country. Incidentally, I picked up some brochures from one of these groups and it mentioned Polyface by name as a hypocrite. The section about us could not have been any more wrong. It was total untruth, written by someone with an ax to grind—some hollow-eyed, narrow-shouldered vegan no doubt who couldn't keep up with me doing meaningful work even for an hour.

I responded that decades ago, when we started, we purchased chicks from a local hatchery. Then it closed. Then we purchased from a farther hatchery. Then it closed. Now we get them out of Ohio or Texas because we can't get them in Virginia. My dream would be to get them from Virginia again, but we need a hundred Polyfaces to create the market demand for that infrastructure to return. Unless and until it does, however, these independent hatcheries and air freight shipping are absolutely the lifeline for the non-industrial poultry movement.

She walked out before I even answered the question. She should have been more honest and asked: "How can you eat chicken when we all know a chicken is the same as a human child?" These animal rights radicals wanting to shut down chick shipping (you have to say that carefully in order to not have a slip of the tongue) are not trying to create chick welfare. They think it's sinful to eat chickens. What's ironic is that they duplicitously fall into a line of thinking that the industrial poultry movement loves: annihilate the small-scale poultry competition. Trust me, the industrial poultry folks love these animal rightists trying to shut down chick shipping. If these radicals succeed, it will destroy the final alternative to Tyson. And wouldn't Tyson love that?

So while they say they are doing this for the chicks, they are actually playing right into a favorite agenda item of the industry: get rid of the pesky independent producers. Before you unleash righteous indignation on something, you'd better be sure about the world such passion will create. The road to hell is paved with good intentions. Righteous indignation is a powerful force. Release it wisely and judiciously, or you might create a worse world than the one you're living in now.

Believe me, Tyson loves it when these animal rightists petition against chick shipping. They probably throw chicken barbeque parties to celebrate their friends at animal rights organizations. True compatriots. Duplicitous do-gooders. Gullible zealots.

Now let's go to the broilers, where Polyface is more vulnerable. There, our farm compromises and uses the industrial double-breasted Nascar race car high octane broiler. Fast growing, heavy breasts. So far, these chickens haven't gone the way of turkeys, where natural service is no longer doable. Chickens still enjoy natural service. Whew! But these chickens still grow unbelievably fast and are temperamental as a result. They are prone to leg problems, heart attacks (their muscle grows too fast for their organs to keep up), and respiratory issues.

Why would we use such an unnatural chicken? Marketability. We tried for several years to offer a non-hybrid bird, but people didn't want it. Marketing is persuasion, and in persuasion you can't push people beyond their tolerance level. If a 1 represents a McDonald's junkie and a 10 represents a true blue foodie with grain mill in kitchen, lard making in slow cooker, local everything and scratch everything, you never try to move a 1 to a 10. You try to move a 1 to a 2 or 3. Otherwise, you're just offensive.

The world will only let you be so weird. You can be a nudist, and you can be a Buddhist, but a nudist Buddhist—that's too weird. So here we are telling people they should buy their chicken somewhere besides Wal-Mart; they should cook it themselves (that's quite a stretch for many these days); they should pay a little more for it. To go beyond that and tell folks not only that, but you also want dark meat, a razor breast, and a chicken too old to fry—you've just become a nudist Buddhist.

Even people who touted themselves as real heritage aficionados didn't want these non-hybrid birds. We used White Plymouth Rock cockerels, 12 weeks old, and called them Marco Pollos (pronounced in Spanish, poyo)—Old Country birds. Here was the shtick: "You may not have been able to sail with Magellan around the cape. You may not have been with Columbus. But you can smell the smells and see the sights, direct from the galley, with the Marco Pollo." It didn't sell. After three years, we discontinued it; simple as that.

What good does it do to go bankrupt being altruistic? At the end of the day, we need to pay our taxes and keep shoes on our feet. Does that mean I wouldn't love to retry this? Of course I would. These birds were healthier, tastier, better grazers. They just don't have a double breast and the meat is tougher and more of it is dark, due to their additional exercise. These are not lethargic birds; they get up and run around.

Does this compromise mean I've joined the dark side? Am I a hypocrite? Have I joined the enemy? Perhaps. I confess it's all a problem. And I don't have all the answers. If we hadn't tried, I'd be more contrite. But we tried. Unless and until we have more people willing to go farther in their 1 to 10 persuasion, we'll produce a bird light years better than industrial but with the double breast that consumers have come to expect. I'll talk more about this in the chapter on pricing.

Hogs. We don't farrow, so we buy piggies from growers in our area who like to do that part. So what do we look for in those genetics? My standard answer: "State of the art 1950s genetics." We want a pig that looks like a pig, not a box. Industrial pigs don't even have a natural gait; they wiggle and walk awkwardly because of the way their rear quarters are shaped. They are too long and too square out their rear end. If you look at any wild animal, their rear end slopes off. It doesn't come straight out and square.

This natural curvature is the phenotype that fosters birthing ease. Gait has everything to do with walking comfort. In the industry, where the hogs never have to move around, gait isn't important. But if the hogs are going to do pigaerating, grazing, and acorn harvesting, they need to enjoy movement. We've tried some of these industrial hothouse pigs, and in our more natural production conditions, they fall apart. They don't gain on foraged feedstuffs like the traditionally-phenotyped pigs.

We want pigs that will put on some fat. Ability to put on fat is a direct indicator of keepability. A hard finisher will take a lot of corn to finish. We want a pig that looks at acorns and gets fat. These are called easy keepers, and they will thrive under more natural (rigorous, perhaps?) conditions. I don't care if the pigs are yellow with pink polka dots. If they work in our protocol, that's all that matters. We've had both heritage breeds and more common breeds.

Generally, on our farm we don't worship breeds. In every breed, you will find animals that thrive under a given set of conditions and ones that won't. The idea is to find the genetic base that works for you. And that will be a completely different base from what will work best in an industrial confinement house.

Cows. We have a small cow herd but we also buy several hundred calves a year from neighbors. We've purchased calves that didn't do well at all under our grass finishing regimen: we don't buy from those farmers any more. We constantly seek and refine, but generally we want a small bodied animal. The industry, because it is geared to corn metabolism, likes huge carcasses. In grass finishing, we want exactly the opposite. We want squatty little barrels standing on toothpicks—again, state-of-the-art 1950s genetics. Easy keepers. We don't want big steam engines that require mountains of feed to stay productive.

This goal, obviously, puts us on the lunatic fringe in the cattle industry. While it's going one way, we're going the other. While the industry awards ribbons to 1,400 pound cows, we're selecting for 900 pound cows. While the industry wants to wean huge calves, we want to wean calves that are half their mother's weight—doesn't matter how big the calf is or how small the cow. Any pair that hits that benchmark is balanced.

The industry wants to grow them as fast as possible, pushing them with corn. We let them grow slower, on perennials. Faster is not profitable. Even race car drivers know that if they push the car faster than its traction, they will wreck. Nothing has unlimited speed. The cost of speed is fragility, more fuel per mile, higher risk.

In the winter, most farmers feed expensive supplements to keep calves gaining fast. At Polyface, we just graze them along on grass and then when it runs out, we feed them hay. The calves grow in frame, but not in muscle. They might even look a little thin. But when spring grass becomes available, our calves will gain up to 4 pounds a day, making up for lost time. This is called compensatory gain, and it is how herbivores throughout the world handle poor grazing conditions. A calf's propensity to gain weight is in direct inverse proportion to its previous 100-day gain cycle.

When farmers feed their calves expensive supplements to keep them gaining nicely in the winter, those calves will actually drop their gain on the spring forage lush. Getting the herd in sync with the solar/perennial forage cycle is the key to profitable and ecological cattle production. While our calves may enter spring looking a bit ragged compared to the ones fed corn silage, a hundred days later ours will be just as big due to compensatory gain. And ours will have achieved that weight at half the cost and energy of their counterparts on other farms. That's sheer ecstasy.

I would be remiss in this whole genetic discussion if I did not introduce every reader to the work of the American Livestock Breeds Conservancy (ALBC). This organization is dedicated to preserving heritage genetics. The goal is to preserve genetic diversity. I deeply appreciate what ALBC has brought to the table and wish more people would use more of these traditional breeds.

That said, let me finish this whole discussion by going off the lunatic edge and introducing you to linebreeding and what I call nativized genetics. Every generation of a plant or animal carries genetic memory that makes subsequent offspring a little more adaptive to the area's ecology. Any Floridian who roomed with a Canadian in college is aware of this adaptation. The Canadian can walk around in a 60 degree room in short sleeves; the Floridian needs an L. L. Bean parka in the same room.

Heritage breeds were developed over many decades, primarily in Europe, and exported to America. These breeds form the basis of what we now call heritage breeds. They were not developed in America. Many of them carry geographic names, like Scottish Highlander cattle or Yorkshire pigs or Suffolk sheep. Farmers in those regions gradually selected the phenotype that proved functional in that area; hence, the geographic names.

I suggest that we should be doing the same things in America. Rather than just preserve Old Country breeds, we need to be developing our own heritage breeds. I think a Swoope cow would be great for my great-grandchildren to enjoy. Essentially, I'm talking about using the same kind of functional selection process used in Europe to give our future farmers a similar legacy. New breeds. Geographically specific adaptive functional phenotypes. How about that?

I confess that I'm not impressed when I meet someone in Alabama effusing about their Scottish Highlander cattle. Why do they have them? Because they're cute. They're different. But these animals were selected for centuries to live in cold, rugged, mountainous conditions. Alabama is practically abusive to them. This is not the way to preserve heritage genetics. They need to be appropriately climatically sited.

So how do we create such a genetic legacy? Linebreeding. Perhaps we could even call it wild breeding. Think about it. Have you ever heard anybody say: "We've got incestuous opportunities out here in the deer population. Goodness, a father might breed his daughter, or a son his mother. We'd better come in here with a helicopter, bring some of those bucks out of there, and move them 50 miles away. Then we'll bring a few bucks from there back here to make sure we have some outcrossing."

No, nobody says that. And yet the wild populations go along just fine with a somewhat haphazard familial breeding program. And they look amazingly similar. That's because functional phenotype and physiology are the only criteria. Not some arbitrary human standard of size, color, or whatever. Over time, this system creates the kind of consistency that's normal in wildlife. Aberrations are extremely rare, and usually don't last long.

Think of how consistently similar deer or squirrels or zebras look. In a herd of zebras, the adult females don't vary by more than a few pounds. The difference between the biggest one and smallest one is practically imperceptible. That's because they've been self-selecting for functionality for a long time. No sophisticated breeder walked in one day and started selecting the ones with the fuzziest tail or the most pointed ears.

Daniel started his rabbits when he was eight years old, as a 4-H project. As of this writing, that's been 20 years ago. For roughly five years, he endured 50 percent mortality. He fed them forage, did not medicate or vaccinate, and did not bring in any additional outside genetics. Close breeding and patience paid off. Gradually, all the maladies began to subside. His rabbits experienced all the problems you see in rabbit rearing books: long teeth, sore hocks, coccidiosis.

But gradually they started to change. Now, 20 years later, without any outside genetics, no medications, no vaccinations, he has the most homogeneous-looking bunch of rabbits I've ever seen. They are cookie-cutter consistent. That's because their appearance is completely functional and not based on anything else. And they haven't been outcrossed, mongrelized, or genetically compromised. Indeed, this is now a Daniel breed, truly nativized.

Fortunately, because rabbits have multiple offspring and a relatively fast generational turnover, Daniel has been able to do this in just 20 years. I figure to do the same thing with our cow herd will take 80 years. Oh well, we're five years into it. But you have to start somewhere.

Some people are calling this wild breeding. Some cattlemen who have adopted this idea are amalgamating their cow herds, running multiple bulls, and not worrying too much about record keeping. If a cow doesn't have a calf, she's culled. They keep bulls from within the herd, from older cows. No cows receive calving assistance. If they die, they die. This hands off approach, while it may sound harsh, or even uncompassionate, is really the most efficacious way to create genetic strength, overall health, and functionality in an environment. Ultimately, I don't have a problem with that.

Well, you say, what about the suffering cow that's trying to calve and finally dies out there? I confess that's not my style, but I completely appreciate the process. Would it have been better to intervene with attention and medication when Daniel was losing 50 percent of his rabbits to birthing problems and sickness? If we had, we would not have the healthy, model group of rabbits we have today. Everyone who sees them exclaims about how healthy and consistent they look. I wish I had a cow herd that looked that way.

The relationship between pain and gain is real: a difficult marital discussion to gain new intimacy; a difficult exam to gain new credentials; a difficult exercise regimen to gain new wellness. You know the Marine slogan: "Pain is weakness leaving the body?" The point is that abundant life is not fluff and fuzzies. Genetic selection is the same way. To really make progress takes some strong culling. That's natural.

What a joy to know that we are encouraging nativized genetics, a true legacy that forms a new heritage for future generations. That's the sheer ecstasy of being a lunatic farmer.

TAKEAWAY POINTS

1. Production per unit eventually becomes inefficient.

2. Radical animal rightists sometimes ignorantly help industrial farming.

3. Breeds known as heritage today were seldom developed in America.

4. Nativized breeds would duplicate the bio-regionally specific functionality that farmers used to develop heritage breeds.

Chapter 16

Artistry and Microsites

M ost modern industrial farming paints with a broad brush. One crop in a big field is normal today. A single land use, up and down the hills, south faces, north faces, east faces and west faces. I never cease to be amazed when I go out to Illinois and Indiana, or Iowa, to see a single corn or soybean field twice as big as our entire farm. It boggles my mind.

Can all that singleness of crop and singleness of land use be good? The single use is a natural result of single purpose. In many cases, the single purpose is commodity farming with subsidies. The narrow subsidy program manifests itself on the landscape with narrow land use. In 1900 these farms were diversified crop, hay, and livestock farms. Today, the fences are gone, the buildings torn down, more than half the population gone, and it's just one plant for miles and miles. Same after same after same. Where's the diversity?

I can hear the industrial farmers screaming back at me: "But haven't you heard about the marvels of Global Positioning Satellites, allowing farmers to customize fertilizer and chemical applications? Now we can change applications from one foot to another in the field based on soil and crop data." Yes, I grant that that is an amazing technology, but it's all based on one crop. In fact, if you change crops, it has no data base.

The result is that this technology is highly linear. If you want to change 5 acres to a pastured poultry operation, the GPS and computer don't tell you anything. Farmers who buy the technology, are more enslaved by its single use than they were before they invested in the technology. The technology itself reduces creativity because farmers feel emotionally and economically dependent on its sphere of knowledge. For some strange reason, the GPS fertilizer data isn't programmed to adjust for bok choy.

This brings our discussion back to the single use capital infrastructure concept. As you invest in single-use technology, you become less and less creative. The single-use infrastructure, whether it's a building, a machine, or a technology, traps you. Instead of it being a servant to your needs, soon you become a servant to it. Divorcing yourself from that marriage is difficult.

I met this reality face to face back in about 1992 when Polyface began raising commercial pastured eggs. We'd been doing the eggmobile, but we needed way more chickens than the eggmobile could hold. At the time, I hadn't conceived of hooking two eggmobiles together. I was locked in this "cleaning up after the cows" paradigm and didn't see the possibility of using eggmobiles solely for eggs. In my thinking, eggs were a byproduct of the biological pasture sanitizer behind the cows. I wasn't thinking about the eggmobile being an egg production model in its own rite.

When we ramped up egg production, then, I simply built more floorless field shelters like the broiler shelters, but tucked nest boxes in the back end. We put 50 layers in each shelter and thought we had the cat's meow. That was when we started the apprenticeship program and when we began selling to restaurants. It all happened in a big whoosh. This model, featured on the front page of *ACRES USA* and other publications, seemed real slick.

A year later Michael Plane, from Australia, came by for a visit and told me about this brand new netting available for poultry. I had never heard of it, had never seen it, had never used it. But I knew it wouldn't work. The reason I'm telling this story is for two reasons: first, confession is good for the soul, and second, as innovative as I am, to be this hardheaded about leaving my old model, I can't imagine how hard it must be for peer dependent, less innovative people.

I listened politely as Michael encouraged me to try it, but I had a million reasons it wouldn't work. The chickens would fly out. Hawks would snatch them. It would be too hard to move. Chickens would get out when you moved it. It wasn't square, so you'd have odd spots around the pasture. Oh, I thought I had it pegged: nope, won't work.

The truth is that I couldn't bring myself to divorce my field shelters. I had built these things, sweated over them, financed them, designed them. I was in love with them, plain and simple. This marriage had nothing to do with what worked. It was simply prideful, selfish pigheadedness. After all, this was my idea. The electrified poultry netting Michael was talking about—that was his idea. Not acceptable.

I dismissed it as unworkable. After all, I had my system. It was a good system. People came from all over the world to take pictures of my system. I wrote about my system. Chefs liked my system. Customers liked my system. Gee, I'm proud of me. I was drunk on pride, pure and simple.

After two years, another friend, Andy Lee, of *CHICKEN TRACTOR* fame, told me I should try this new electrified poultry netting. I thought Andy was pretty clever, but I didn't think he was as good as me. Maybe a better writer, but definitely not as good a farmer. My old competitive pride kicked in and I decided to try it just to prove to him that it wouldn't work. I wanted to say "I told you so."

So I got a piece of this newfangled netting and encircled one of the field shelters, propped the shelter up on a 5 gallon bucket, and let the chickens out. They loved it. I know a happy chicken when I see her, and the hens were ecstatic. They ate more grass. The ate less grain. Predators did not come. The hens stayed in—they didn't even attempt to fly over the netting or jump over it, even though they could have done both. I was amazed.

Gradually my old Grinch heart began to soften. Within a week, I announced: "This is it. This is the new paradigm. Within a year, the individual shelters will be abandoned." The marriage was over. The divorce was complete. I had a new lover—electrified poultry net. When I decide to divorce an old protocol, I do it completely and quickly—one of my strengths, I think.

We first built an A-frame prototype field shelter on skids for about 400 birds and it worked quite well. Then we upgraded to a hoop structure on skids. It worked very well. Now we have a scissor-truss structure that works better. And who knows what the next 10 years will bring forth? The point is to stay flexible, hold your infrastructure loosely, and always be ready to jump to a better ship.

Very simply, if I may digress to national policy for a moment—since truth is truth and everything relates to everything—this is what's wrong with the national fuel alcohol program. Not that alcohol fuel is a bad thing. But when you build subsidized single-use capital-intensive multi-million dollar alcohol plants, the culture does not abandon them easily. The truth is that whether or not we need the energy, whether or not producing the corn to fuel those plants is ecologically sound, whether or not those plants are economically viable, those plants will still operate because the culture invested time and emotion in them. And the more you have invested in the relationship, the harder it is to break it off when it's no longer healthy.

Those alcohol plants will absolutely dominate all landscape decisions radiating out however far is necessary to keep raw ingredients flowing into them. I don't know if it's 50 miles or 100 miles. But the existence of those plants will become the dominant decision-making force for how the earthworms are treated in the entire region. That's the nature of the beast. The same holds true for silos, CAFOs, and combines. The bigger the infrastructure, the less flexible. It's much easier to turn a speedboat than an aircraft carrier.

The answer for energy is to eliminate the Bureau of Alcohol, Tobacco, and Firearms (BATF) and let anybody who wants to have a still in their backyard have one. That's the way it used to be, before prohibition. If you research it a little, prohibition did to decentralized and independent energy production what the drug wars have done to decentralized and independent food production. A government that has the authority to criminalize your backyard whiskey production can and will also criminalize your backyard energy production. And a government that has the authority to criminalize drugs can and will also criminalize raw milk and compost grown tomatoes.

All under the guise of safety and the common welfare. What a joke. More about that in the next chapter. Just a bit of foreshadowing.

Back to the broad landscape brush. Whether it's massive strawberry fields in California, massive corn fields in Iowa, or massive chicken factories in Arkansas, scale is important. Things that are appropriate on one scale do not necessarily stay appropriate on a large scale. Think about a great teacher you've had. If she's a great teacher, why don't we put everybody in her class, so everyone can benefit from her skills? What if we make that memorable 25-person class 100 students instead? How about 500? Now do we have the same thing? Of course not, and lots of things in life are that way.

But Wall Street acts as if hockey stick lines are wonderful and nothing to worry about. As if hockey stick trajectories can go on indefinitely. I've got news for you: they don't. And they won't. Actually, the big picture is the sum of lots of little pictures. Landscapes are the same way. The more we appreciate and leverage the subtle nuances of a landscape, the more productive, more stable, and more sustainable it will be.

With that in mind, I like to think that here at Polyface we do not paint with a broad brush. I want a tiny brush. I want to tease out every possibility, both in the landscape and in the people on our team. Just like people have various gifts and talents, the landscape does too. Finding and capitalizing on them creates a mosaic of symbiosis and synergy.

A wet weather seep can be dug out, dammed up, and developed into a year-round water reservoir. A ravine can be dammed up, like we talked about in the water chapter. The point here is that most farmers are not even thinking about micro-sites. Goodness, mention chickens in a National Cattleman's Beef Association (NCBA) meeting and they will call security and have you arrested—or as close to it as they can come. Cattleman hate poultry. They blame poultry for chipping away at their market share. But it never occurs to them to quit feeding herbivores like chickens. They just hate chickens.

Instead of leveraging the cow's unique ability to convert perennial forages into nutrient density, thereby doing something the omnivore can't, cattle farmers feed the same thing chicken farmers and hog farmers do. That loses the unique advantage cattle enjoy. The digestive tracts of animals are some of the most unique micro-sites on a farm, and appreciating their differences is key to capitalizing on their strengths.

I don't think a farm can ever fully utilize all the micro-site nuances within the farmstead. A few years ago we were creating a new two acre pasture in an area that had been plowed 200 years ago and in about 1920 was abandoned and grew back up in forest. Since I'm the main chainsaw operator, I was cutting along and came upon a small sugar maple, perhaps 4 inches in diameter. Our farm is in the edge of sugar maple country. We have a couple of sugar maples in the yard that Daniel has tapped occasionally since he was a child.

Rather than cutting that sugar maple (after all, we're making a field here so ideally we'd minimize trees) I left it. I was glad I did because a couple of days later, I found another one just like it. In that project, I found five young sugar maples that now form a neat line right through that little two acre field. Because the cows are only in there about four days a year, the trees are growing nicely. The cows haven't destroyed them. In fact, the cows shade up under them and add urine and dung, which are helping the trees grow much better than they were when in the forest and shaded by the overstory of more dominant trees.

Since that time, we've found another four sugar maples in the same area. Two of them are already big enough to tap. I realize Polyface is not in the maple syrup business. I don't know if we ever will be. But by appreciating those trees, we are now only a decade or so away from a 10-tree sugar bush. Not commercial, by any means, but certainly enough to produce several gallons of that sweet elixir. From our own farm. With our own hands. When the Commie-Pinkos come, or the Greenies, or the Religious Right—whatever, we will have a sugar bush. And to me, that's valuable—emotionally, if not economically.

Compare that to the timber harvest contract we executed a couple of decades ago when we traded 30 acres of forest for 3 miles of all weather road. I'm not at all sorry we did that. It revolutionized the farm and gave us access to 300 acres. But in that contract, it specified that everything larger than an inch in diameter either had to be taken away or cut and left. I wonder if there were any fledgling sugar maples in those areas? But big outfits can't do the kind of individualized tree-by-tree careful assessment that I can do because they have to keep a skidder, knuckle boom loader, and a couple of trucks busy. Since I don't have all that infrastructure pushing me, I can study the trees, sit and think a bit, study along and be far more sensitive and aware.

What a wonderful dividend when that level of observation pays off in a sugar bush. Many years ago I was cutting off a one acre ridge and I encountered a tree I'd never seen before. I wasn't sure what it was so I left it. It was a pretty good size, probably 12 inches in diameter and straight as an arrow. Very pretty. I figured it would be nice to have more of this tree, whatever it was.

When the leaves came out in the spring, I got my tree book, and (I know, you readers in Colorado are going to laugh at this) determined it was an aspen. For the record, we don't have aspens in this part of the world. I was pretty excited. I had an aspen tree. What I hadn't seen was a big squirrel hole about halfway up this 50 foot tree. The next winter, with all the trees removed from around it, a windstorm broke off the top at that squirrel hole. But since I had opened up the area around it, new aspen trees sprouted that could not have sprouted as long as the heavy forest was there. Now we have an honest-to-goodness aspen grove in that area. It's probably 50 feet in diameter with maybe 30 nice saplings. I'm sure some if not all are root suckers, but they are all growing well and many are now 20 feet tall. Very pretty in the fall.

What's the value of an aspen thicket in Virginia's Shenandoah Valley? I have no idea. But it's pretty. And it's unique. There again, a commercial logging outfit would have rolled right through there and chopped it down. Who knows where that initial seed came from? Clearly, something on that site is friendly to aspen growth. So I'm enjoying it, along with the personal satisfaction of that unusual circumstance right here on our little farm. Isn't that cool?

One of the concepts in permaculture is to examine a piece of land for a year or two before doing anything. Walk it once a week, or at least once a month—notice I said, walk it, not drive it.

Make notes about what you see and what you feel. For example, air flows in rivers just like water. For almost half a mile, at the bottom of one of our fields and along the edge of a long flat one, I've always noticed a warm tunnel of air in the spring.

You can walk into and out of these air streams. They are most noticeable during the spring and fall, when the weather is changing and a bit more unsettled. Of course, a warm air stream keeps buds from frosting. As we've worked in this area, we've found several old apple trees in that line. They follow right in the center of that warm air tunnel. Clearly, folks long ago felt that stream too and planted some apple trees there.

By walking the property, you find out where the seeps are. These would be places to avoid with a road or building. Maybe a wet spot is a potential pond site. You would find out what areas dry out sooner. That's where you want buildings and roads. If you see wild grape vines entwining some trees, that means the site and soil are good for domestic grapes too. Perhaps that's the area for a vineyard.

Too often farmers purchase a piece of property and never ask it what it would like. I know some of you may be thinking I've really gone off into la-la land. You're thinking: "Now he's showing his true colors. What do you mean, communing with the land?" I'm not talking about an audible conversation. But humbly moving across it, observing, listening, feeling, touching. Where does the snow pile up? Where does the honeysuckle grow? Where do groundhogs like to dig? That probably indicates deep soil. Where do walnuts sprout? Where does the sun hit first in the morning? That might be a good place for a greenhouse or solarium on a dwelling. Where does the sun hit last in the afternoon?

At the risk of going off on a tangent, let me relay a real story that speaks to this depth of understanding. Hugh Lovel, guru of biodynamics in Georgia, wanted me to install one of his cosmic energy accumulator pipes on the farm. I've been asked to try lots of things over the years by innovators hoping that I'll love something, endorse it, and tell everyone else to use it.

For the uninitiated, a cosmic pipe is a 10 foot piece of PVC buried 2 feet in the ground, oriented due north, with two T's plumbed in to hold a baby food jar wrapped in copper wire. The jar contains customized homeopathic remedies. The claim is that the pipe equalizes energy fields between the atmosphere and the earth, promising: "you'll never have another drought." Sounds good to me.

I asked Hugh where the pipe should be placed: "At the convergence of the energy fields on your farm, and somewhere the cows can't rub against it." I picked a spot—I mean, a square foot spot—for this thing to be. A month later Hugh came with the materials for the installation. He wanted to walk around and see the farm before we dug the hole and planted the pipe. He brought his dousing rods (some people use a pendulum) to determine the best spot to place the pipe.

When he got out of the car, he took a dousing rod bearing to get a direction and we set off on a circuitous walking tour around the farm to see the animals and pasture. Periodically, he would stop and get another bearing with the dousing rods. I did not tell him anything about the site I had picked. He was going strictly on the dousing rod directions. As the sun began to set, I got worried that we wouldn't have enough time for the installation before dark, so we decided to get earnest about finishing the project.

He took a bearing and we were off. I didn't let on anything about the direction we needed to go to find my location. He followed his dousing rods—in a beeline toward my location. My heart began to pound. I didn't say a word. We got within about 20 feet, and he stopped: "We're real close." I couldn't believe it. He began slowly circling, watching the dousing rods, moving closer and closer to my location. In about 5 minutes he stopped and announced: "This is the spot. All the energies converge right here." Are you ready for this? He was standing on the exact square foot I had picked intuitiviely.

I couldn't believe it. He had never been on our farm. Had never even seen a picture of it. He was following dousing rods and I was going on blind intuition. I began laughing, and explained that we had both picked the same spot. He concluded that I was so tuned into this farm, that it just subconsciously moved my spirit. Sounds good to me. I know Native Americans reading this will not think this weird at all. Only Westerners who want to reduce everything to nuts and bolts will find it a little flaky. But this is the level of love, observation, and care that creates a climate of ecstasy. To the average farmer, this kind of talk is just lunacy.

We installed the pipe. The next summer we had the worst drought we'd had in 100 years. Okay, apparently the dousing rods work better than the cosmic pipe. If I had dismissed the whole attempt outright, how boring and humdrum my life would be. Now I have this marvelous, fun story to tell, all because I was willing to engage in lunacy.

Landscape micro-site observation precedes infrastructure development. Once you move in and start developing the farm, new opportunities arrive. For example, roof eaves are a wonderful high moisture area for growing things. The drip line off of a roof edge offers lots of extra water. If the roof is metal, it often drips condensation from dew in the summer. A raised bed of vegetables under such spots can leverage that extra water.

On the south side of one of our barns, along the drip line, we plant cucumbers every year. Since this is adjacent to winter cow and pig housing where we do the deep bedding, some of the bedding compost squeezes through the retention wall and falls out under the drip line. The extra moisture and the compost combine to create a conducive habitat for red wiggler earthworms. They come into that roughly two foot wide zone, all along the drip line. You can dig down with your hands and pick up handfuls of red wigglers. It's wonderful. In that zone, then, we plant cucumbers. The cucumbers climb up the retaining logs and fencing, enjoy the rich worm casting soil, and the extra moisture from the drip line. It's a perfect micro-site.

We've even rolled round bale feeders into that area, turned them on edge, dropped twine down from the top, and trellised the cucumbers. What do you do with round bale feeders in the summer? We use these feeders as portable feeders in the winter, even though we don't use round bales. They are good to feed any kind of hay. But they are never used in the summer. Cucumber trellises are a great use, and create artistry in what otherwise would be just a weedy barn edge.

Shiny metal siding reflects sunlight. By using shiny metal for the sides of buildings, you can create warm micro-climates on the south side of outbuildings. Yes, colored siding may look nicer to you, but think about the advantages of creating a long area that warms up a month earlier in the spring and stays warm a month later in the fall. That's not something to sneeze at. If you put a cold frame there, you get double solar gain. How would you like head lettuce in the winter? Sounds good to me. Perhaps this site would be just warm enough that you could have a fig tree.

A similar idea uses ponds for thermal mass. In our area, breezes come from the west. As the air blows across a pond surface, it picks up the stored heat in the water. The water cools and heats much slower than air, so the water holds a more constant temperature. The leeward side of a pond, then, offers a wonderful micro-site for frost-sensitive plants. If the morning is real cold, the vapor created as the cold air hits the water surface, then wafts gently across, acts as a high moisture fogger for the leeward side plants. This artificial fog reduces frost risk because the cold air freezes the moisture instead of the plant leaves.

Another permutation on this theme is the floating garden. I got this idea out of the Farm Show magazine and we are still developing it. It's not cheap, but it does work. We purchased 20 pieces of 10 foot X 6 inch diameter PVC, capped the ends, and lashed them together to form a raft. Two pipe pieces went up the edges to keep the raft pieces from moving out of their plane. Filling the raft with compost and soil, we planted vegetables on top.

The plants grow in soil but their roots can go in between the PVC tubes to be as wet as they want to be. The plant gets to pick how wet its feet are. The pond creates an insect moat. Fish eat the bugs as they try to get to the garden. The water acts as a heat stabilizer so that on the hottest summer day, the water cools the plants, and on a cool spring or fall day, the water warms the plants. The water reflects sunlight to the underside of the leaves, encouraging faster growth since sunlight is hitting on both sides of the leaves. The raft shades the pond. The aquatic life enjoys the shade. The roots offer a high oxygen zone to the water, and the rootlets offer a special critter habitat for the fish and frogs to nibble on. And it's beautiful.

This kind of careful site leverage and resource use can never be done on a wide brush, massive scale. But if leveraged fully at the micro-site level, the amount of symbiotic production that a small area can achieve is practically limitless. I always wanted to put a screened floating bucket on the pond with chicken guts in it that would attract flies. The subsequent maggots would fall out through the screen and feed the fish below. If I ever do this, I'll call it the Grub Tub.

When you start down this path, the opportunities are limitless. All it takes is thinking small, thinking relationally, and thinking customized. Think of a tiny artist's brush as opposed to a spray painter. I'd like to think that the Polyface farmstead is a detailed canvas as opposed to an industrial farm, which is a spray painted parking garage.

Don't forget terraces. I think some of the most amazing landscapes in the world are terraces. Talk about taking low productivity sites and turning them into Edens. Not only do terraces reduce flooding by slowing down the water, they create much more surface area on a slope. By stair stepping the slope, you double the surface area. That in itself creates more area to collect solar energy. If the land surface is the mete and bound of solar biomass conversion, then certainly doubling the land surface increases solar collection. How cool is that.

But it's hard to build terraces with big machinery. All of the famous terraces in the world were built by hand. They are farmed by hand. And many formed the backbone of some of the world's greatest civilizations. I think terraces are kind of like ponds—you can never have too many. That means you can spend a lifetime building them and enjoying them. Now I'm sounding like something is limitless. Well, ponds and terraces come as close to limitless as anything I can imagine.

J. Russell Smith's *TREE CROPS* set the standard for multi-tiered farming. Carefully selecting complementary canopy types, he developed systems using, for example, grape vines, then low growing apple trees, then higher growing nut trees. On the ground, you could run livestock or poultry on the forage. That creates four tiers of production on one site. How many tiers are in a corn field? Are you starting to get the picture?

When you begin thinking like this, the ideas start to flow like water out of a pipe. The permutations on the symbiotic, micro-site, stacking concepts are endless. The whole process has its foundation, though, in individualizing, customizing, landscape plans. It's all about appreciating site differences rather than trying to create sameness, like single-crop industrial systems.

When I start one of these brainstorming sessions, I can hardly go to sleep at night for all the micro-site opportunities. The mosaic that such development creates is both functional and beautiful. Truly, it's the sheer ecstasy of being a lunatic farmer.

TAKEAWAY POINTS

1. Let the landscape speak.

2. Listen to the landscape.

3. Capitalize on artificial micro-sites created by infrastructure development.

4. Wait and observe before plunging ahead.

Chapter 17

Honest Pricing

Every time I speak somewhere, I can be assured of being asked two questions:

The first one is: "That's all fine and dandy, but can this system actually feed the world?"

The second one is: "How can we make this food affordable?"

We've dealt with question one in great detail, except for one nuance: nobody goes hungry due to a shortage of food. America throws away some 40 percent of all food produced. I was out in Idaho a few years ago and went to a thousand-cow dairy farm feeding potatoes. Apparently these spuds weren't quite the right fit for fries.

The world produces way more than enough food to feed everyone on the planet. Half the food imported to India is consumed by sacred rats. That's not a food production problem. The inequities in food, and world hunger, have nothing to do with enough food being produced. The problem is distribution. Neither an industrial system, organic system, beyond organic system or

214

anything else from the production end can necessarily change the distribution inequities.

When a tribal chief or warload won't let a Red Cross truck pass into a starving African village, that's not a production problem. When free U.S. foreign aid food dumped in a village displaces the indigenous food system and drives all the farmers out of business, that's not a production problem. For the record, nobody is starving because there isn't enough food. The world is not overpopulated. Distribution is a completely different problem from production.

However, if production were more localized, distribution would be less of an issue. The answer to localized production is not western food dumping on hurting nations. It is the kind of farming espoused in this book—lunatic farming. And that will not be encouraged by USAid, the Peace Corps, UNICEF, the United Nations, the World Health Organization, Bill Gates, or any of the rest of the big charitable or governmental aid bureaucracies.

I spoke in Arkansas a couple of years ago and a group of military personnel came in, dressed in their fatigues. After the talk, I went over to them and talked to the Lieutenant Colonel in charge. The group was being deployed to Afghanistan to do agricultural rural development. The Lieutenant Colonel said: "Your talk was great. This is exactly what we need to do in Afghanistan. But you know what we're going to do? We're going to give them equipment they don't need that will require fuel they can't afford that will break down and need parts they don't have to grow crops they don't eat to be fertilized with U.S. subsidized chemicals on land that shouldn't be plowed...." I can't remember the whole monologue, but it was long and passionate.

Then he said: "You know what these people need? They need to feed themselves. They need root cellars, backyard poultry, village-scale processing equipment, earthworm beds, greenhouses, electric fence, ponds, and portable infrastructure—just like you just showed us on this slide program. Would you be willing to come and talk to some generals?" I assured him I would be glad to. Never heard from him again. Can you imagine how he was treated by the higher ups? These alleged military heroes are just pawns providing corporate welfare to American industrial agri-business. That's all it is. And it's despicable. Enough already. World hunger can only be solved by local production. It can never be solved by foreign aid, no matter how well intentioned.

The food deserts in inner cities, where cigars and Ho-Ho cakes are available but a whole tomato or whole chicken can't be found do not exist because tomatoes or chickens are in short supply. And they certainly don't exist because of racism. Yes, these areas have a lot of minorities, but that's not why supermarkets don't want to locate there. These areas are also starved for clothing stores, dry cleaners, auto shops, art galleries. They are what they are because the culture that exists there is barbaric.

Barbaric cultures don't attract meaningful commerce. It doesn't matter whether the people are white, black, purple or blue. If they act like animals, respectable people don't want to go there. That is why I'm so excited by inner city gardens. Vacant lots can be used for production. And every time vacant land is converted to food production, it tames down the neighborhood. It builds pride and sense of worth. It fosters a can-do spirit and radiates around the community. These areas will never be healed from the outside in. They will only heal from the inside out. When leaders step forward with courage and vision and refuse to be barbarians, it brings order out of chaos, vision out of victimhood, and dialogue out of discord.

Hunger, whether it's American inner city, rural Appalachia, former Soviet Bloc or Africa, is always best solved from the inside out, locally.

"Can you really feed the world?" Yes, and here's why.

1. Scientific ecological farming is only as old as scientific chemical farming. If you visit any living history farm circa 1900, you will not see a compost pile. That's because modern scientific composting was not widely used until Sir Albert Howard did his trials in India during the 1920s and 1930s, subsequently popularizing the nitrogen, carbon, oxygen, moisture, and microbial formulas in what many see as the beginning of the ecological farming movement. His AN AGRICULTURAL TESTAMENT was first printed in 1943.

Remember, worldwide soil depletion, desertification, and land degradation predates chemical and industrial agriculture by a long shot. The American answer to soil degradation was simply to move west.   When Americans ran out of west, and began urbanizing during the industrial revolution, many people began studying soil restoration. Although Howard may have been the big dog in that effort, he was by no means alone.

After World War II, chemical fertilizers beat out the composters due to several unfair advantages:

a.   Ammonium nitrate, super triple phosphate and other chemical fertilizer formulations were the same as ammunition and explosives formulations—well proven, well known science and easier to learn than composting.

b.   Bombs had already paid for the chemical manufacturing infrastructure, so the true cost of these fertilizers never expressed itself on the price tag. The military industrial complex capitalized chemical farming.

c.   Composting required biomass pulverizing and transport, which had not yet been perfected.

d.   Advertising inertia favored the chemical companies, who ended the war with stashes of cash to leverage on a duplicitous public.

e.   Bombs are sexy;   compost isn't--although more sex happens in a compost pile than in an explosion. As the industrial revolution permeated the national psyche, American culture embraced factories, manufacturing, and store bought. Even breast feeding fell into disrepute for a couple of decades until the back-to-the-land mother earth revivals of the early 1970s.   Free love eventually trumped bombs.

2.   The infrastructure and scientific understanding developing around ecological agriculture paralleled the chemical approach in magnitude and time.   Industrial food advocates consistently rail against Polyface that our practices are a return to hog cholera, poultry Newcastle's disease, brucellosis in cattle, and tuberculosis in humans, as if our farming represents the epitome of Luddite mentality.

Nothing could be further from the truth. The urbanization occurring in the 1920s and 1930s predated electrification, refrigeration, stainless steel, sewage systems, and basic sanitation knowledge and practice. Innovation never occurs simultaneously along all the ancillary edges of the change. It's a ragged edge. The point of the innovation always extends far beyond the support infrastructure thought and hardware to fully metabolize the innovation. Related innovation takes awhile—like a slinky effect.

Farm labor migrated to cities by the hundreds of thousands during this urbanization, pushing farmers to embrace industrial practices before electrification, stainless steel, refrigeration, pharmaceuticals, nutrient cycling, building design and machinery developed to metabolize the new industrial farming conditions. This lag created mud lots and hog cholera epidemics around the nation. Ditto for dirty dairy. Ditto for poultry diseases. Routinely feeding antibiotics to farm animals in order to keep them alive in crowded mono-species conditions was still two decades away.

Meanwhile, in the quiet revolution occurring at Malabar Farm in Ohio (Louis Bromfield), the Rodale Research Center in Pennsylvania, Ed Faulkner's trials (Plowman's Folly), William Albrecht in Missouri and other giants of the ecological farming movement, the infrastructure and understanding to complement Howard's composting innovations were gaining ground. Efficient chippers to reduce biomass into decomposable and easy-to-handle pieces became widely available.

Hydraulics finally made their way onto farms in the early 1960s. By the 1970s, 4-wheel drive tractors were available, which made hydraulic front end loaders affordable and efficient even for a small farm.

Meanwhile, electric fence came of age in New Zealand during the early 1970s. What had been a cumbersome and undependable innovation became highly dependable, energy efficient, and incredibly portable. For the first time in human history, large scale commercial herds and flocks could be controlled efficiently to mimic the movement patterns of massive natural flocks and herds. Cation exchange capacity, magnetized foliar feeding, and a host of other earth-shattering developments occurred in this renegade world.

But the culture was fixated on irradiation, genetically modified organisms, DDT, Agent Orange, oxytetracyclene and the techno-glitzy innovations coming from the chemical-industrial paradigm. The Polyface paradigm was shunned like an ugly stepsister. You didn't read about it on the front page of the New York Times.

From 15-year UV-stabilized canvas covers, extruded steel

tubing for hoophouses, meticulous planting and harvesting machinery to our own pigaerator compost innovations utilizing symbiosis and synergism, the innovation and high-tech natural solutions to food production were just as profound—and certainly less risky—than the highly publicized chemical-industrial discoveries. So when the industrial food advocates accuse Polyface of wanting to return to hog cholera, it's disingenuous in the extreme—nothing could be further from the truth. They assume that while the chemical-industrial system innovated, the ecologically-sensitive system remained static. That's ridiculous. Polyface is not Grandpa's farm. Anyone visiting Polyface will see, in just a few minutes, a dozen high tech innovations Grandpa could not even have imagined.

And the truth is that if the same time, energy, and creativity invested in chemical-industrial models had been leveraged on composting, chipping, and portable infrastructure, America would be producing far more food today than chemicalized mono-cultures, with more nutrient density, building soil instead of continuing to erode it, without a Rhode Island-sized dead zone in the Gulf of Mexico, and would not have poisoned eagles, frogs, and salamanders. That was a long sentence. Now catch your breath. Bottom line: if America had conducted a Manhatten Project for composting, the race would have been far different.

3. Western research does not measure whole systems. The naysayers from the United Nations and corporatized research institutions like land grant colleges measure only one component when they study indigenous, diversified food production systems. They only measure rice production; they don't measure rice plus ducks, plus duck eggs, plus talapia, plus arugula and bok choy. The fact is that these highly choreographed symbiotic systems produce more food per acre, in aggregate, than the most heavily fertilized genetically modified rice because to produce that rice, the paddy is too toxic to support ducks, fish, and salad greens.

This kind of compartmentalized, agenda driven research permeates countless official findings and government reports. This junk science finds a home every day in the media and the minds of duplicitous people.

4. Contrary to popular thought, Polyface pastured systems do not take one iota more land than Tyson factory chicken

houses—or any other Concentrated Animal Feeding Operation (CAFO) for that matter. The alleged land efficiency of a CAFO is a charade. What do you think those animals eat? They eat grain. And where do you think the grain is grown? Somewhere. Maybe not there, but somewhere. The point is that those CAFOs are not stand-alone entities. Imagine, extending out from each one, acres and acres, even square miles, of subsidized annual grain production.

At Polyface, the omnivores that do eat grain substitute a portion of it with perennial salad bars, so if anything, our production model requires less land than CAFOs when all is said and done. And beyond that, perennials thrive on land that would not be suitable for tillage. This difference opens up countless more acreage to the food production pool. The bottom line is that empirically, Polyface produces more food per acre than industrial-chemical systems, and is in fact the most efficacious way to feed the world. That reality is sheer ecstasy.

But what about question number two? How to make local food affordable? Everyone knows how high farmers' market prices are. Organics. Whole Foods. What do we do about the price issue? This question has a tremendous amount of unspoken assumptions behind it. Let me itemize some of the assumptions to set the context for the discussion:

1. Organic is for elitist blue hairs.
2. Local farmers are charging an exorbitant price and getting rich.
3. If it's local, it doesn't have to travel as far so it should be cheaper.
4. I can't afford it.
5. If I spend that kind of money, I can't buy a flat screen TV.
6. The poorest person deserves this food.
7. Taxpayers should subsidize local food.
8. I'm a college student, and I don't have time or money.

You get the drift. It's all pretty damning. What's funny about this whole discussion for me is that Polyface has been accused for many years, by the organic community, of being too low priced. In a spirit of full discloser, I'm not a pure capitalist. I don't believe in charging what the market will bear. I believe in making a profit and being efficient. Beyond that, a professional salary is plenty. What in the world would I do if I earned a six figure income? Who needs all that money? I like being poor because it denies taxes to the government. Anything that keeps those lousy politicians from getting their grubby greedy hands on my money is good, including poverty. Most of my life, as far as the government's concerned, I've been below the poverty level. I enjoy being an American in poverty. Now do you still trust those poverty figures?

When Teresa and I started out, we lived quite comfortably on an income a third of the official poverty level. We grew all our own food, heated with our own firewood, wore cheap and second-hand clothes. We lived in an illegally refurbished attic apartment above my parents in the old farmhouse, never went out to eat, had (and still don't have) no TV and enjoyed NPR's radio classics on Saturday nights for entertainment. I always said if we could figure out how to grow Kleenex and toilet paper, we could pull the plug on society. Though the government said we were poorer than poor, I guarantee you we were much happier than most millionaires. And still are.

Dad always talked about an elderly aunt who frequently prayed for "those poor rich people." Why not try to live as cheaply as possible, instead of as expensively as possible? Do you know how much fun it is to live cheaply? I guarantee you won't worry nearly as much as people who are trying to hold onto their money. They have to worry about the stock market, Wall Street, and interest rates. The more insular your living requirements, the less vulnerable you are to all the headline paranoia. If I could design a perfect world, I think it would exist without money. Okay, back to the point of high priced food. The one exception to all this is that we do spend extra money on good food.

Here at Polyface, we put a value on our time, like $25 an hour, and then work back to a pricing that gives us that return to labor. A pure capitalist will generally charge whatever the market will bear. But I'd rather charge as little as necessary to return a professional salary, and then figure out how to meet the demand if the market grows. The alternative is to stay at the same level of production by raising the price to curb demand. Probably a hybrid of the two is a more realistic approach.

Beyond that, I don't try to have the most elite product. We'll resume here the discussion of the Marco Pollo chicken. Because that chicken grew slower, we had twice as many pastured shelter move days—twice as much production labor in the birds. Since they were older and more exercised, they were tougher and therefore much harder to process. In the time we could do 100 pounds of 8 week double breasted broilers, we could only do 50 pounds of the Marco Pollos. When the tissue is tougher, everything from knife cuts to evisceration to tucking the legs in at the end is harder.

Double the labor in production. Double the labor in processing. And when we were done, instead of a carcass that averaged 4 pounds, we only had one that averaged 3.5 pounds. The only part of this bird that was cheaper was the initial chick cost. And that was frankly part of the allure for me, because hatcheries just kill most of the cockerels since the demand is for the pullets. As a result, these non-hybrid cockerel chicks— remember we used White Plymouth Rocks—only cost about 23 cents apiece.

All in all, though, we needed nearly double the price of the regular industrial Cornish Cross to make the same return to labor. That put the price up to $5 a pound. A retailer, marking it up yet another 30 percent, would sell it for $6.50 a pound. That's expensive chicken. We'd actually have a better chance getting that for our double breasted industrial birds than these heritage breed birds.

I really believe it would be great to completely supplant the industrial food system with pastured poultry. I'd love to see the day when not a single CAFO operated. Realistically, in order to do that, we can't charge $6.50 a pound for chicken. Here's a valid question: is it better to produce a perfect product and sell it at a price that only 1 percent of the population can afford, or produce a 90 percent perfect product that 99 percent of the population can afford?

That, by the way, is exactly the question Steve Ells, founder of Chipotle Mexican Grill restaurants, asked. And that is why he stuck with his restaurant model. If you talk to Steve, he will tell you that the secret to Chipotle's success is good sourcing (we supply the pork for two of their restaurants in Virginia—

Charlottesville and Harrisonburg), traditional cooking techniques, and the open kitchen. Normal fast food in America violates every one of these protocols. Certainly Chipotle can be faulted for some things, but I've been impressed at the creativity and constant progressive push. Always trying to do better.

By far and away our most competitively priced product is our salad bar beef. The reason is because it's the most scalable. A herd of cows requires a front fence, back fence, water trough, and mineral box whether it's 50 or 500 individuals. If you're going to move them every day, the cost of being there is the same.

The next most competitively priced product is pork, partly because of scale similar to the cows, but also partly because the pigs are picking up a large portion of their diet cheaply. Self-harvested acorns and roots, for example, are much cheaper than purchased and transported grains. Those feed savings, we pass along to the buyer. Pastured rabbit is the next most competitive product for the same reason. By reducing their purchased feed costs 75 percent, we can raise a much better rabbit at less cost. Although those feed costs are offset by higher labor costs, the net is that we are extremely competitive to industrial retail prices.

The poultry is a different story. First, it's not as scalable. The 98th dozen eggs requires the same handling to put away as the 97th dozen. Ditto processing the 98th broiler vs. processing the 97th. Industrial poultry is far more scalable because sophisticated machinery works cheaper by the million, and drugs substiture for labor in the production phase. That is why our poultry and eggs carry a higher sticker price, compared to industrial supermarket fare, than the beef and pork. Automatic egg processing equipment and evisceration machinery is extremely expensive. Only large establishments can process enough material to justify these capital intensive machines.

Now let's broaden this discussion. Local foods generally, and Polyface foods specifically, carry a higher sticker price for several reasons. First and foremost, we receive no subsidies. We take no government money. Period. We don't take grants. I don't even know where the offices of the federal agencies are located. I don't use them, I don't want them, I don't need them. As far as I'm concerned, the USDA, Small Business Administration, or whatever, don't even exist.

I'm not asking vegans to pay for my poultry processing facility. That's what happens, you know, when tax free industrial development bonds sweeten investment deals for Tyson processing plants. I don't understand why Americans worship Abraham Lincoln—he gave us the USDA, for crying out loud. He was the first and most aggressive big governmenter, along with Teddy Roosevelt (food police), Woodrow Wilson (income tax), Franklin Roosevelt (social security), Lyndon Johnson (welfare), and Barack Obama (health care). Isn't it interesting that the most famous American presidents are the ones who flouted the Constitution? The ones who adhered to the Constitution have largely been forgotten.

Corporate welfare comes in many shapes and sizes. I have to make my own way, tote my own satchel, figure out my own financing, and pay front end money for everything. And most of the time, I take it on the chin for being non-industrial. For example, under workmen's compensation, I can't have a low risk delivery driver. I can have only a live animal hauler because Polyface is a farm, not a delivery business. Our interns and apprentices have to be registered for either cattle or poultry. According to workmen's comp rules, people who work with cattle cannot work with poultry. Is that insane, or what? To haggle through all of this, I have to sit here with auditors and insurance agents and figure it out. That takes valuable time that could be spent producing food; instead, I'm haranguing with a bunch of bureaucrats. Somebody has to pay for that time—guess who pays?

If the U.S. had four automobile companies and decided to subsidize three of them at $5,000 a vehicle, but not the fourth, who would have the lower prices? Wouldn't it be crazy for people to jump on that fourth car company, alleging: "It's not fair! You're an elitist, charging those prices." No, what's unfair is the subsidies. That's exactly what we have in the local food system. We haven't entered that privileged class where we're too big to fail. We're too small to be noticed. Instead of bail outs, we're bullied by bureaucrats because we don't have the political clout to get them fired if they harass us.

The second reason local food carries a higher sticker price is due to regulations. I've already talked about this, but I'm going to push forward with it a little bit here. Again, if you want to know a lot more about this, you can read all the stories in my book *EVERYTHING I WANT TO DO IS ILLEGAL*. Let's take bacon as an example. At $9 a pound as of this writing, that's expensive. Let's break it down:

$1.50 a pound for processing. We talked about food policy, economies of scale, and why local abattoirs are at a tremendous cost disadvantage.

$2.00 for curing. We have to take it to a federal inspected curing facility. Because they really don't like to do this small batch work, they charge an arm and a leg for the service.

$1.00 for packing. Again, that's charged by the curing facility.

$1.00 for putting the label on. Charged by the curing facility—the only legal facility in Virginia, by the way.

We're at $5.50 right now and we haven't yet bought or birthed the piggie, paid for the feed, paid to transport the finished hog to the abattoir or picked up the cured product. We haven't yet paid for any infrastructure or any production labor. We haven't paid for any marketing or overhead. Folks, at $9 a pound, we're losing money! All of this nonsense could be done right here on the farm for pennies. Instead, we burn up a whole truckload of fuel running up and down the interstate, transporting it hither and yon for this part and that part, and at the end of the day, we haven't made squat.

By the way, this is NOT sheer ecstasy. It's sheer madness. The only time our kind of farming is negative is when we interact with bureaucrats. I'd like to see them all fired and get real jobs, the kind that actually produce something instead of being parasites. Let them make payroll, fill out reams of paperwork and wake up every day fearing that they've missed a dot on an I or not crossed the top of a T and are ready to face the wrath of bureaucrats and gun-toting law enforcement officials "just doing their jobs." Many horrible, terrible things have been done in history by people just doing their jobs. I don't care what job you take, you're still a human, and last I checked, humans are supposed to have a heart and a conscience. Apparently bureaucrats check theirs at the door when they go to work.

I'm convinced government agencies have conscience scanners at the door to the office building. "Sorry, ma'am, you can't take that conscience in there. Have to leave it with us. You can pick it up this afternoon when you leave the building. Excuse me, sir, you have reason. No, that's not allowed today. We're unable to be reasonable here. You have to leave reason behind. Prejudice? Oh, yes, come right on in. No problem. Lots of that here. Special concessions to big players? Ah, we love you. Please come on in."

Here's another example of capricious regulatory costs. It's called the 30-month rule. Since Mad Cow disease has not yet been found in an animal under 30 months, the USDA, in its pontifical wisdom, has decided that beef animals more than 30 months old need their backbones removed, to get rid of the spinal cord (known as Specified Risk Material—SRMs). According to the USDA, mad cow is caused by cows eating dead cows.

A grass fattened animal, therefore, by the USDA's own definition, cannot get mad cow. And mad cow isn't communicable in the air. Wouldn't you think Polyface could get an exemption to this rule since we don't feed dead cows to cows? Not on your life. Why would we want an exemption? Our beeves grow slower than the ones fed corn and steroids. Sometimes it takes ours 30 months to finish. If we lose the backbone, we can't have T-bone steaks and some other high end cuts. And the abattoir paperwork is so horrendous that we pay an extra fee. It hurts us two ways. In France, by the way, 36 month old heifferettes are the premium beef delicacy.

A calf's teeth come in just like a human child's. The USDA has determined that if the second group is in, then the animal is 30 months old. Arbitrarily. You and I both know that all children get their teeth and lose their teeth at a certain day of age. No differences. I mean, you can just mark it on the calendar that the first tooth will come in at 156 days of age and a child will lose the first one at 1,234 days of age. Of course not. Some come in early and some late. Want to know a little secret? Cows are like that. Are you surprised?

One of the components of tender beef is early pubescence. Farmers who grass finish select, genetically, for early pubescent cattle. One little problem. You know those 30-month teeth? Calves can get them as early as 26 months. So here we are, the only true antidote to mad cow, being penalized for selecting grass-friendly genetics by the industrial food police. These malicious, capricious, asinine regulations drive the cost of local food through the roof. If it weren't for these, local food would spin circles around industrial fare—in price and quality.

The third reason Polyface products have a higher sticker price has to do with externalized costs. That fecal pall that hangs over the Colorado feed yard and sickens the workers and nearby community is not paid for by the consumer. Those recalls for pathogens and food borne illnesses aren't paid for by the consumer. The 1,000 tractor trailer loads of turkeys allegedly destroyed in Virginia during an avian influenza outbreak a few years ago were indemnified by the taxpayers. How about the Rhode Island-sized deal zone in the Gulf of Mexico?

I could go on and on, but the fish kills from Smithfield manure lagoon outbreaks, repetitive motion illnesses, tainted water and stinky air—none of this is on the sticker price of industrial food. All the cases of food borne bacteria-induced diarrhea. What's a case of diarrhea worth? More than half comes from tainted food. All of this is acute stuff. How about the slow moving stuff like Type II diabetes, heart disease, and other maladies yet to be determined due to nutrient-deficient foods?

A caveat right here. Perhaps you think I'm making all this up, and you're saying: "Okay, Salatin, where's the evidence? Cite some empirical examples." If that's your attitude right now, I could list a hundred specific examples, I could quote the Centers for Disease Control, I could probably quote the Virgin Mary but you wouldn't believe it. You see, our heart defines what our heads will believe. The old saying: "I'll believe it when I see it" is actually backwards.

The truth is: "I'll see it when I believe it." I have never convinced anybody about the evils of non-industrial food. Or the righteousness of honest food. Never. Every person comes when their heart is nudged toward the truth. I can't think of any more powerful story to prove my point than Easter. Jesus was put in the tomb, a heavy stone rolled over the entrance, and a heavily armed, elite Roman guard stationed to secure it. When the angel came and the guards fell down as dead men, the angel rolled the stone away and Jesus came out of the tomb. The guards staggered back to headquarters and the cultural leaders paid them money to say: "Somebody came and stole the body."

Dear reader, if elite Roman solders, a hand picked security force to guard the tomb, who then witnessed the resurrection of Jesus, could be bribed to say it didn't happen, then people can be bought off, consciously or subconsciously, to sell out for any lie you can imagine. Go ahead. Think of any lie. Anything. Perhaps in our times the only thing that could approach this is the person who says the Holocaust was a hoax.

I am incredibly grateful, and blessed, to know that I'm not killing anybody with my food. I'm not hurting people. I'm not participating in a lie. I'm offering honest food at an honest price. All the costs are figured in. I'm not destroying people's lives and debilitating their bodies in anti-human working conditions. I'm building immune systems, not destroying them by drugging their dinners. What a relief. Thank you, Lord.

Now that we've examined the major reasons our sticker price is higher, I want to deal with the affordability issue. How do you know something is affordable or not? I think this is a lot like the "I don't have time" excuse. I like to finish the time excuse with "for that." That's really what it comes down to. We tend to make time for what we consider a priority. Remember when I cooked the omelets for the college students? Who drank a soda? Who watched a movie? I'm not in the victim business.

The same is true with the word affordable. Just so we get down to brass tacks, let me itemize a few things that nobody has to buy, ever:

1. Fast food.
2. TV
3. Movies
4. Soda pop
5. Alcoholic beverage—beer, wine, liquor
6. Cigarettes, drugs, etc.
7. Diapers—use cloth ones
8. Baby food—grind regular food with a table top grinder and let the feeding begin
9. Vacations—stay home and read a book and play games
10. New cars—buy used and save

11. New clothes—buy at the thrift store.
12. Processed food –buy raw and prepare it yourself.
13. Junk food—snacks, potato chips, crackers
14. Breakfast cereal—fry an egg or make your own granola
15. Candy—chocolate, sugary anything
16. Eating out—my apologies to all my chefs. I know most people won't get this serious, so no need to worry. Teresa and I didn't eat out at all for our first 5 years of marriage. We were poor and at home I ate like a king from our own pastures and garden.
17. Recreation—bowling, ice skating, etc. Have fun at home.
18. Gadgets—let the Chinese keep their trinkets.
19. Toys—kids generally like boxes more than toys anyway. Be creative.
20. Furniture—you don't need anything fancy.
21. Cell phones
22. Entertainment centers
23. Music
24. Life insurance
25. $100 designer jeans with holes already in the knees
26. People Magazine
27. Hotel rooms for $8,500 a night

I could go on, but a list like this helps to put some things in perspective. We are a pampered, self-serving, materialistic, spoiled brat of a culture, in my opinion. And if you can afford these things, go ahead. But don't come whining to me saying "I can't afford your food" if you're participating in any of this. Are you serious about eating well, or not? Let's cut to the chase and quit pussy-footing around: if your lifestyle includes any of this stuff, I don't believe your whining. Period.

If we really believe things, we will do what it takes to make them happen. Otherwise we're just playing games. All of us, including me, can do much more than we're doing. I don't mind people not doing all they can do. But be honest about it and don't complain to me that my food is not affordable. If you extricate yourself from the above list, I guarantee you my food will be quite affordable. And your quality of life will increase, too.

I really believe that if we took all the money we spend on junk and pseudo-food and converted it to honest food, plenty of money exists in the system for everyone to eat royalty quality food. Don't tell me you can't afford good food while you're smoking on a cigarette or chugging a beer. The only thing worse than being uncompassionate to real hurts and needs is being all goo-goo eyed over someone too lazy or undisciplined to get with it. Misplaced charity to whiners is as inappropriate as withholding charity to real needs.

Finally, our food carries a higher sticker price because it's worth more. Flat out. No question. Forget arguing with me. It's worth more. You can measure the nutrition empirically. You can look at the farm's ecology both intuitively and empirically. You can taste it. You can smell it. You can touch it. You can feel it in your body. This isn't some hyped up sales pitch. We've heard from too many customers who explain, in all sorts of ways, how normal food is superior. Remember, industrial food is abnormal.

Routinely, normal food farmers collect testimonials about the superiority of what they produce. I don't care if it's wine, vegetables, or meat, local and non-industrial is worth more, just like a Rolls Royce is worth more than a Volkswagon. Does anybody waggle their fingers at Rolls Royce and say it's unfair for them to price their car higher than a Volkswagon? After all, they are both just cars. They both have just one engine. They both have only four tires. But you and I both know they aren't the same.

I assure you that for a host of reasons, both emotional and empirical, pastured chicken is a whole different critter than industrial chicken. Compost-grown local tomatoes aren't even in the same league with cardboard genetic gas-ripened trucked-in tomatoes. According to the USDA, a tomato is a tomato and a chicken is a chicken. It's all the same. Generic nutrition labels are all available. But those labels would be turned upside down if the USDA actually started checking normal food.

When we had one of our chickens tested at the Virginia Tech food sciences lab, the PhD in charge of the tests handed the test back to us and said: "See, I told you there wouldn't be any difference." When we opened it up, our fat profiles were so different from industrial that the instrumentation line almost went clear off the sheet of paper. The percentage differences were up there at 100 and 200 percent. This isn't the same stuff. It's not anywhere close. But the expert was so immersed in the paradigm that a chicken is a chicken is a chicken, he couldn't see the evidence to the contrary even though the instrument needles impaled him in the nose. That's how powerful paradigms are.

Think about it. What would the average American have to give up to eat truly honestly priced, normal local food? Candy? A couple of movies a year? When you think about it, not much. It's an attitudinal thing, not a financial thing. Let's be honest about it. The money is in the system.

No discussion of pricing is complete without including the difference between processed and unprocessed food. Our culture has this universal perception that processed food is cheap. Frozen microwavable pizza, supposedly, is cheaper than food at the farmers' market.

Well now wait a minute. If you purchased all those frozen pizza ingredients in a raw state and made them in your own kitchen, it would not be more expensive. Abdicating kitchen responsibilities and turning them over to processing factories does not make for cheap food.

For example, a pound of premium grass-finished Polyface ground beef costs less than a happy meal. Anybody want to compare nutrition?

Although I have never done this, I am confident that if you took the average shopper's grocery cart and added up the cost of all the processed goods, they would be far more expensive than all those items made from scratch ingredients bought unprocessed. The more processed the item, the more true this is. All those emulsifiers, stabilizers, and items you can't pronounce don't enter the meal cheaply.

To be fair, I'm not putting a price on the kitchen labor. That's true. But I'm assuming the nutritional jump, the social equity, and the spiritual satisfaction compensate for that. Recently in Australia I was prepping to go on their Today show. The show prep lieutenant threw me some questions to see how I would handle them. She kept trying to get me to agree that Australia needed more regulations to keep processors from processing so much.

Regardless of how she framed the question, I remained adamant: "We don't need regulations. Nobody is making people buy processed food. The responsibility for what to buy and what to eat lies with the individual, not the government. If people abdicate their food choice responsibilities, that's not the government's problem."

By this time, things were getting a little hot and heavy in our interchange, and she responded: "I think our viewers will be offended if you tell them they are responsible."

There is your conventional media, folks. People are stupid; the government is our nanny. Believing in personal responsibility is just lunacy. Didn't that go out with the Dark Ages? Come on, join the Twenty First Century, where we all know the government is the answer to everything.

You can buy a 10-pound bag of potatoes for the cost of half a pound of potato chips. You can buy fruit juice and carbonated water for soft drinks. Who needs frozen pizza dough? You can buy a bag of flour for pennies compared to industrially processed material. People who consistently buy unprocessed can buy first class ingredients and stay competitively priced with junk ingredients turned into processed foods.

Of course, this requires domestic culinary skills. But that's not a food price problem. That's an education problem. It's a want-to problem. Let's not call price a food issue when it's really a get-off-your-duff-and-get-in-the-kitchen issue. I know this is not politically correct, but somebody needs to say it. Just for the record.

Who needs boxed cereal? Teresa gets raw ingredients, mixes them together, and then bakes them in the oven on a pie plate to make homemade granola good enough to die for. The bottom line on this hearty breakfast cereal is much cheaper than processed boxed varieties. And I love her more for it. What's that worth?

For one last blast on this issue, did you know that for about $8 billion every school lunch in America could serve honestly priced local food? When you think of the nearly trillion dollars given to the scalawags in banking and finance in the bailout plan initiated by former president George W. Bush and perpetuated by president Barack (Change) Obama, you can't help but think the culture is just like the government. For literally pennies compared to the wars in Afghanistan and Iraq, and the giveaway bailout to crooks in corporate high places, every school lunch could have royalty food. But no, that's too hard to do. We've got to throw the money down a rathole instead. It does more good there, you know.

Affordable? I think so. Our government mirrors the priorities of its people. Affordable is a subjective priority. All we lack is the will, the intestinal fortitude to make honestly priced food a necessity. Being able to look any neighbor and any customer in the eye and know that my food is honestly priced is part of the sheer ecstasy of being a lunatic farmer.

TAKEAWAY POINTS

1. The world has plenty of food, and will for a long, long time.

2. Hunger is due to distribution problems, which can only be solved by local food systems.

3. Industrial food is dishonestly priced.

4. Plenty of money exists for every person to eat like royalty—it's all a matter of priorities.

Promote Community

Chapter 18

White Collar Farmer

One of my favorite salary-related stories is from a guy who was a candidate to be a pastor in a church. Located in an upscale community, the board of elders consisted of white collar professionals. One was a doctor, one an attorney, one an accountant and two were bankers.

During the interview process, the elders always asked prospects about their salary expectations. Ministers, of course, taught to be deprecating and non-aggressive, always deferred to the budget: "Whatever you have in the budget will be fine with me." As the selection process wore on, they finally had a sharp candidate come who really had passion and talent. Impressed with his style and heart, they hoped he would consent to come.

The interview proceeded apace and all seemed well. Subconsciously, each of the elders had already picked this fellow. He endeared himself to them long before they came to the salary question. Thinking this a rather routine part of the interview, they awaited his answer: "I'll be happy to receive the same amount as the average of all your salaries."

236

The room fell silent. Stunned elders lowered their heads and studied the floor. That was not what they had expected. After a pregnant pause while the elders inhaled and rubbed their hands, the candidate continued: "That's not what I demand, but if you aren't willing to do that, don't you think it says something about the value you're placing on your spiritual condition?" The elders, cut to the heart, realized that they had viewed spiritual growth far differently than economic growth. Quickly they approved the candidate, with salary as the candidate requested, and the church grew spiritually far more than before.

This story is a great backdrop to one of my great passions: white collar farmers. Perhaps this is why I'm captivated by the agrarian gentility of our nation's founders. By extension, southern plantation gentry has the same attraction. Don't misread this: I'm not yearning for slavery or anything close to it. But I deeply appreciate a culture in which agrarian vocations are not only considered acceptable, but worthy of the best and brightest in the land.

We could quibble about timing, but I think especially since the Civil War, farmers as a class have been marginalized. The best and brightest in any culture tend to gravitate toward vocations of greatest remuneration. I could argue, nobly I think, the fallacy of chasing remuneration. Wouldn't it be great if no Chief Executive Officer took a greater salary than three times the lowest salary in his business? Talk about endearing yourself to the worker bees.

Certainly we could wax eloquent about non-monetary values. I've touched on that already in this book. Three-legged salamanders are more important than business plans and return on investment. That's true. And we should all aspire to noble and sacred vocations first; money second...or third or fourth or fifth. The point is, a lot of things are more important than money. Point duly noted.

All that said, however, I actually believe we need professional farmers. Smart farmers. The best and brightest. And to do that, farmers need to believe, first of all, that such a possibility exists. The poor dumb farmer stereotype is ubiquitous in our culture. I remember like yesterday when my high school guidance counselor learned I wanted to be a farmer: "What? Throw your life away?" The poor woman went apoplectic. I mean, we all know that honors graduates and debate trophy winners don't become farmers.

Two couples in a spotless BMW drove up to our farm sales building one day. Clearly well bred blue hair country club types with quaffed hair and stylish clothing, they came into the sales building to buy some things. I was in there with two of our apprentices and one of the ladies began conversing with us. After about five minutes, she got this startled look on her face and said: "You guys are so articulate. Why in the world would you want to be farmers?"

I smiled broadly and in genteel southern affability accentuated with appropriate hillbilly drawl responded: "Oh, we're pretty good about covering up our stupidity." Then I spat on the floor—no, not really. Let's dissect the layers of assumptions in this question, because it has lessons for us.

1. Farming is not for the intelligent. Let me get this right. Stated another way: food production should be the responsibility of dolts. Oh, that's it. We want the least intelligent to grow our food. Now let's see, would we apply that to anything else?

Probably not. Not even our plumber. For the life of me I can't figure out why people think idiots should be in charge of their food. We don't even think that poorly of clerics. What kind of food is likely to be grown by the kind of person you think should grow it? Have you ever wondered why farmers make such stupid decisions? Like, knowing all we know today, still signing up to become a slave to Tyson to grow chickens? I mean, after all the terrible things that we know about how the company jerks the farmer around, the farmers who have lost their farms and been destroyed by these companies. And farmers are still signing up. Dumb, dumb, dumb.

The cultural assumption becomes the reality. The more and the longer a culture creates a perception, the closer reality comes to that perception. If farmers are supposed to be the dumbest class in society, who do you think will ultimately dominate that vocation? The perception becomes a self-fulfilling wish. How many farmers read books? Not very many. The ones who do are lunatics through and through.

But if you look back at the farmers who framed our nation, they had extensive libraries. They were men of letters. The South developed the academy. Planters had their faults, but education was not one of them. As a culture, we will not have intelligent farming until we have intelligent farmers. And we will not have intelligent farmers until we believe intelligent people can and should farm.

2. Farming is not for the articulate. By and large, the founders of this nation who debated at a level that today's college graduates can't even follow, honed their skills reading and talking about significant issues first around their dining room tables, and then in their communities.

Farmers were the CEOs of that era. Just like the movers and shakers of modern America come from the ranks of corporate CEOs, in the 1750s farmers comprised that class. Not all of them, by any means. But good farm CEOs were looked up to in the culture. No stigma was attached. If one of these guys walked into a public meeting, he wasn't assumed to be a stammering idiot because he was a farmer.

When people ask me what course of study in further education an aspiring farmer should pursue, my standard answer is "learn to communicate." If you can communicate you can do anything. But that's not part of agricultural degrees. Why would a farmer need a speech class? Tractors, cows, and manure spreaders don't talk, so why should a farmer? In fact, most agriculture degrees are not for people wanting to be farmers, but for people going into agri-business to tell farmers what to do. That's the great misnomer about agriculture degrees.

They teach the Latin names of all the bones in a horse, but don't teach how to know a horse. The ag degree folks who eventually do end up being good farmers spend the rest of their lives de-learning what they spent big money to learn. I had a recent land grant graduate in animal science spend a morning here on the farm and when he left, he said he learned more in three hours than he had in four years at college.

If we actually placed articulate farmers on a pedestal, and compensated them with tangible and intangible rewards, the ranks would swell with articulate farmers. If I started a college for farmers, here is how I would set it up. First, I'd search for the best and most articulate farmers to be mentors, to be the faculty. I'd look for vintners, horse farmers, cattlemen, horticulturalists, orchardists, apiarists. You get the idea.

Each mentor would define a time period for the topic, number of students, and remuneration requirement. Students would select their curriculum from the mentor listing. Some modules would no doubt be a month and others would be much longer. Students would pay tuition just like at any regular college, but by the end of four years would have had a dozen or so varied and valuable experiences.

The college staff would help the mentors create their curriculum and design research papers and/or testing criteria. The mentors would do what they do best and the college would offer backup academic support. Students would have these wonderfully varied experiences being tutored by the best masters in their respective fields. This model would offer unprecedented exposure diversity, and beautifully blend theory and practice, cerebral and physical. A degree from a college experience like this would insure common sense and practical understanding.

3. Farming is undignified. This is not about intellect, nor about articulation savvy, but about class. Do classy people farm? That is the question. When asked: "Name some classy people," how many farmers come to mind? Classy people are the kind of people we want to be around. Well mannered, conversational, hospitable, clean, and stylish.

An apprentice recently said that a significant part of my success is that I dress and look like a professional. Many years ago I began wearing a coat and tie to do speaking engagements because I saw myself as a professional. I'm embarrassed when farmers show up with their hair topsy turvy looking like something that just blew in from the barnyard, wearing patched overalls, faded shirt, and worn out shoes. That may be the attire on the farm, and that's fine and appropriate. But when Mr. Smith goes to Washington, as the movie by that name suggests, you should look like you belong there.

My years of debate experience taught me that dressing sharp not only helps win the argument, it also shows a respect to the audience. Think about it. If I walked in to do a talk in my farm attire, the persona would be this: "Gather 'round, ever'body. We're gonna have a sideshow of country bumptkinism." Contrast that with when I walk in wearing a tailored suit: "We're here to participate in a credible professional discussion about changing the world." See the difference?

Recently, as I've done more higher profile media appearances, I don't know how many hosts have expressed their dismay: "Aw, you didn't wear your hat." Or, "aw, I thought you'd wear your farm clothes." I know these hosts mean no harm by it, but it always strikes me as being a bit condescending. What do they want, a hillbilly farmer act, or a classy agrarian professor? I'll take the latter, thank you. We've got way too many of the former.

Most of the farmers I know seem to love being unclassy. Kind of like a pig in the wallow. This whole societal persona and expectation makes farmers look at themselves as unclassy. Once again, the cultural perception becomes self-actualizing, and that's a shame. I'm just loony enough to think that farmers should listen to something besides country music. That they should feed their minds something besides NASCAR and John Deere sales brochures.

For class, the genteel southern planter was without peer. Their writings are sprinkled with quotations of classic English poetry and passages from mythology and ancient Greek literature. At Polyface, our apprentices eat their evening meal with us, and Teresa teaches proper etiquette to these bohemians, many of whom have never learned about proper utensil placement. And yes, we pass the food. Everything is supposed to be passed to the right. Come on, people, get with the program. If we're going to grow royalty food and eat royalty food, we should exhibit royalty graces when we dine.

If you ask for the potatoes, you don't just grab the spoon and start dipping while your neighbor sits there holding the bowl. That's uncouth. We say grace before meals. This creates order and decorum. It's not just pigs seeing who can dive in first. We wait and defer to the hostess to give the go-ahead. We want our apprentices to leave knowing these social graces because we want to turn out classy farmers.

4. Farmers are supposed to be poor. As you can imagine, this assumption really grates on my nerves. In our culture, even janitors are supposed to have a vacation. Are we willing to pay

enough for food to enable farmers vacations? I'm not talking about those socialist anti-liberty farmers that take government subsidies. I'm talking about the kind of farmers who show up at farmers' markets. The kind of farmers that drive a local food system.

After our previous discussion about honest pricing, I hope it's obvious that good farmers need to be rewarded for their effort. Particularly good farmers. Farming as a profession is not even recognized any more without a vow of poverty.

One of our former apprentices from Washington state fell in love with a Canadian, right across the border, and had to fill out Immigration and Naturalization Service (INS) papers to get married. The INS doesn't want Americans marrying foreigners only to join welfare rolls. That's commendable.

What's not commendable is that when he filled out his occupation as "Farmer" they kicked the paperwork back, saying that farming was not an acceptable occupation. He had to agree to a viable occupation for five years. So he started driving trucks. What does it say about a culture when farming isn't even considered a viable vocation? He'd have probably been fine if he registered as a bank robber. Come to think of it, maybe they assumed that "Farmer" meant he was going to be a welfare parasite, since so many of them are. And they just didn't want two.

Depending on who you read, farmers receive between 10 and 18 cents out of every retail food dollar. The old saying that the cereal box costs more than the farmer receives for the corn in the flakes is very true. In other words, if the USDA decided on a new agriculture policy: "Farmers will no longer be paid for anything. All of them work for nothing," that policy would only change the cost of food about 15 percent. That's not much.

Carlo Petrini, founder of Slow Food, invented the term co-producers for people who don't farm. He was trying to help people understand that whether or not you grow food, you either participate directly or by proxy. How you eat and what you buy determines the kind of farming we have. Hence his idea of co-producers. Not bad.

To drive that point home, let me ask a question. What would you think if your farmer showed up at farmers' market driving a Mercedes Benz? Think about that for a little bit. This is a really telling question because it gets right to the heart of the matter. I don't want to get into a debate about whether or not somebody should drive an exotic, expensive car. It's just a metaphor for luxury, and you can see it a lot easier than a Caribbean cruise or an African photo-safari or a host of other things considered the domain of the upper crust.

The quality of your diet will be in direct proportion to the value you place on your farmer. If you think farmers should be paupers, you'll never eat the food white collar farmers produce and you'll always whine and complain that royalty food is unaffordable. If, on the other hand, you think it would be a great thing to have your farmer show up in a Mercedes Benz, you'll enjoy royalty food and actively participate in the landscape ripple effect that such an attitude engenders. Thoughtful farmers create thoughtful landscapes.

Here at Polyface, we see the whole spectrum. Plenty of people say "you're just farmers. I can get it for a lot less at Wal-Mart." That always grates, as you can imagine. But others say: "Oh, that's too cheap. You should charge more." Guess which one of these is fostering a cultural climate of professional farmers instead of unthinking dolts who just do what corporate officials tell them?

Here's a benchmark for knowing we'll attract the best and brightest. The next time you find yourself at a soccer game and all the soccer moms are huddled over at the edge of the field prancing around with their chests stuck out extolling their little virtuosos and their stellar career possibilities, and one of those moms says proudly: "Well, my Mary (or Billy) is going to be a farmer," and everyone breaks out into applause, then, dear friends, you will know that better food and better land stewardship are here.

Frequently I'll have an apprentice or intern come to me discreetly and say: "Mom and Dad are having some trouble with me being here. They helped pay for this computer degree...or business...or whatever and here I am talking about farming, and they aren't too happy about it." What's disturbing is that more often than not, these parents are members of Nature Conservancy or Sierra Club or some other name brand environmentalist organization, but they are still under the dumb farmer spell.

I spoke one spring at the annual regional shindig of Nature Conservancy (NC). Wonderful people. NC was trying to protect special habitats in the region. These were unique ecosystems with rare plants and rare animals: some unique feature. They were acquiring these either by easement or outright purchase. A map showing the locations was color coded, one color for these areas, one for private farmland, one for municipal land and one for government land.

Since this was in the mid-Atlantic region, by far and away the largest portion of land was privately owned farmland. While I applauded the protection effort, I made the point that if all that private farmland became an ecological disaster, those islands of uniqueness could not be protected. This was the point Allan Savory made to me at one of our first encounters: "You're not sustainable." I was offended. How dare he say that about me?

But he made the brilliant, and accurate point that our farm cannot be an island. Its hydrology, rain cycle, soil, and other elements are directly dependent on neighbors' farming methods. If theirs turn into desert, nothing we can do on our little piece will keep it from turning into desert. This is the nature of connectedness. Allan was not saying I was doing bad things; he was just pointing out that no person, no farm, no environment can be disconnected from its surroundings.

My challenge to the Nature Conservancy folks, then, was that they needed to patronize good land stewards in the area of those special lands. This is the boots-on-the-ground reality so many fail to make. I realize all organizations have their focus, but too often the mission itself becomes disconnected from the bigger issue. Possibly, if not probably, these critically identified areas would be much better served by Nature Conservancy if it aggressively discouraged grandparent members from taking their grandchildren to McDonald's than by litigating easements or acquisitions on these isolated rare landscapes.

Sometimes I wonder if a lot of members join these groups for guilt assuagement rather than real engaged cultural alteration. The thinking is like this: "Now that I've given my money to this outfit, I can check off my environmental commitment for the year and keep eating candy, junk and soda, and taking the grandkids to junk food joints." This is all part of that disconnectedness in our culture created by compartmentalized thinking.

This is my problem with carbon trading and cap and trade schemes. What is this, earth stewardship by indulgence? As long as I can pay enough into the treasury, I can live like the Devil any time I please and be forgiven? Just keep the money rolling in? Eventually, somebody needs to just do what's right instead of getting permission to do what's wrong. I've heard this called: "doing bad but feeling good about it."

If all the money funneled into these big name environmental organizations were channeled into identifying and then turning the membership loose on ecologically beneficial farmers, it would do a lot more planetary healing than litigating, regulating, and legislating. They could start their own subsidy chest so that if school districts couldn't afford ecologically-friendly food, the environmental organization would throw in a subsidy. Can you imagine the impact that would have?

Then the message would be localized. The battle would be fought over the dinner plates of real people discussing real issues in their homes and classrooms instead of artificial news conferences and lobby rooms in bureaucracies somewhere. If all the money and time spent regulating the word organic had been spent freeing up local food entrepreneurs, we'd be spinning circles around industrial food by now. Instead, we squandered all that investment fighting ourselves, dividing ourselves, while the industrial food system expanded and laughed.

Here's the bottom line: if you want good farmers, then focus on farmers. Don't go chasing after ancillary things. If you want good food, focus on farmers. If you want good land stewardship, focus on good farmers. Farms control more of the landscape than anyone else. Doesn't it make sense to populate the landscape with good farmers?

To attract good farmers, they must be paid. They must be professionals. They must earn a white collar salary. Lots of them don't deserve a penny. Bad farmers should go out of business. Nobody should buy from bad farmers. They should either be terminated or cut out of the market. Some people think I put too much emphasis on the business side of farming, or on making a profit farming.

What's wrong with profit? It's the lifeblood of business. It's the lifeblood of sustainability. How sustainable is an unprofitable farm? And I think if you want to farm, you should do it. And you should do it fulltime. And you should expect to make a white collar salary. I don't apologize for making money. It's not the only goal, or even the greatest goal, but we'll get a lot better farmers incentivizing them with a white collar salary than we will trying to incentivize them with a vow of poverty.

As a culture, we need to dig down deep into our collective psyche and ask ourselves: "What vocation can we least do without? What's our most important career?" Is it accounting? Is it heart surgery? Is it teaching? How about software development? I'm certainly not trying to trivialize every other vocation. But I think until we realize how few really good farmers we have, and how valuable they are, we will not have good land stewardship, good food, or healthy economies.

An economy can only be as healthy as its farmers. Farmers drive the lion's share of landscape stewardship. Ultimately, if the landscape ecology fails, the economy will fail. Farmers drive food quality. Ultimately, if health fails, the economy will fail. The reason I'm beating this issue is because I find myself fighting the "just a farmer" mentality. I get on an airplane and the seatmate smiles and asks: "What do you do?"

"Oh, I'm just a farmer." Please forgive me, Lord, for answering that way. No, no, no, a thousand times no. How about this instead: "I'm a professional land healer super nutrition food purveyor landscape architect nurturer." I guarantee you that will get a response: "Come again?"

I've spent some time here carping on non-farmers...oops, co-producers trying to create white collar farmers. The other side of the coin is how farmers perceive themselves. Do I even respect myself? Do I wish I had done something else? Too many farmers do. The reason I don't chop prices down to poverty status is because I respect myself. It's not prideful to say that I deserve to earn a decent salary.

That's just down home self respect. I'm not a trashy farmer, and therefore I don't deserve a trashy salary. How we view ourselves actually creates the person we become. If I don't think I deserve a decent salary for being an exceptionally good farmer, guess what? I'll probably never get a decent salary. A decent salary begins with my perception of me. That's the way it is in any vocation.

My Dad always joked about how ministers were always called to bigger churches. Ever notice that? Most say God calls them to their place of ministry. But isn't it interesting that God always calls them to bigger pulpits? Why doesn't God ever call a successful minister down to a struggling, flailing group that's as poor as a church mouse? It could be greed. It could be self-indulgence. And it could be that God appreciates incentive too.

I never want to be wealthy, and I love giving money away. But I don't know another vocation where people are expected to work the number of hours that farmers work, without getting paid for it. We routinely put in 90 hour weeks. What would a white collar salary putting in that kind of time be worth? Forget salary, how about just a wage earner—with that amount of overtime? And yet if a farmer would happen to drive around in a Mercedes Benz, people would think he didn't deserve it.

Farmers should aspire to make a good living on their land. They should aspire to be able to go out and eat once in awhile. Maybe even go on a vacation. Maybe even buy an expensive cowboy hat. Unfortunately, farmers typically feel like they've betrayed their vocation if they splurge on some self-indulgence. If you're a good farmer, healing land, nurturing eaters, you jolly well deserve to be pampered by society. And I say shame on our society for failing to do so. We have exactly the kind of farming and food that we deserve. By withdrawing white collar professional status from the vocation, we've cheapened nobility and adulterated sacredness.

I will never apologize for believing good farmers should be compensated well. Wouldn't we have a better culture if excellent farmers received as much as excellent heart surgeons? What if we had such good farmers that heart surgeons became obsolete? What if we had such good farmers that we didn't need school nurses anymore? What if we had such good farmers we didn't need chemical companies anymore? What if we had such good farmers we didn't need feedlots, CAFOs and pesticides anymore? If all the wealth going to CEOs of these detrimental or remediation businesses had been channeled into good farmers, what a different society we would have.

The reason I can't and won't sell chicken for 89 cents a pound is because it insults my profession and the earthworms under my stewardship. I refuse to join the cheap food dumb farmer cultural agenda. The fact that the industrial food system does sell it for that, and pats itself on the back for doing so, simply illustrates the lack of respect in the whole system. That farmers voluntarily rush to participate, to join in with such a disrespectful system simply shows their duplicity. Truly thinking, innovative, entrepreneurial farmers would not fall for the industrial temptation.

I enjoy holding my head high as a farmer. Not just a farmer. A farmer in the Jeffersonian model. Businessman, professional, man of letters and lover of discourse. Why am I so unusual? I should be normal. Completely normal. If I hadn't thought I could make a good living on this farm, all the altruism and ecology and beauty would not have drawn me here. I'd like to think I'm not mercenary, but all of us have a price. My price for farming was the ability to make a decent living.

Dad was an accountant and had a lot of farmer clients. I never knew who most of them were, and he certainly never divulged names with stories. But he would routinely lament the general poor economics of the way they handled their farms. On our farm, he went the opposite direction he saw most of them going. Small or no machinery. Portable instead of stationary buildings. Pasture-based rather than grain based. Perennials instead of annuals. Seasonal instead of year-round. Carbon cycling instead of petroleum based fertilizer inputs. Direct marketing instead of wholesale commodity marketing. All of this, of course, was total lunatic thinking.

Seeing how hard farmers worked for such little pay had a profound effect on him. When he and I would talk about options for me, if it didn't return a decent amount, he would grin and say: "You might as well do nothing for nothing as something for nothing."

Believe it or not, many farmers would be more profitable if they quit and did nothing. If you're showing a loss every year, you'll come much closer to profitability if you shut down and go to zero. It might not be positive, but at least it's not negative.

Cutting expenses is important. But at some point you have to generate income. And believing that a decent income is not only proper but possible is absolutely the key to attracting the best and brightest to farming. Our culture has designated A and B students as non-farmers and C, D, and F students as farmers long enough that the dream doesn't even exist in the minds of sharp kids. Imagine if our kids were building symbiotic, synergistic ecology-healing farms in their minds like we encourage them to build spaceships or computers? What a different world we'd live in.

Although astronauts aren't drawn to that vocation for the money, if it doomed them to a life of poverty, it would be hard to make that vision shine. The life of an astronaut isn't shabby. A good farmer should live as prosperously.

Countless times, people have said to me: "I always wanted to farm, but I didn't think it was possible to make a living. If I had only known it could be done, I would have taken Grandpa's farm. But now it's gone. The family sold it a few years ago." You can see the lower lip tremble. Moist eyes. Faraway look. What could have been. What could have been.

How else can you explain all the 50-year olds jumping into farming? Burned out in their Dilbert cubicle. Tired of working for someone else. Desperate to grow something, build something tangible. My guess is that this desire was there when they were 10, 13, 16, and 19. But it could never surface. It had to be pushed down. Because farming was just not a valid vocation. The wasted creativity our culture has suffered due to this perception borders on criminal. We've wasted sharp minds that could have focused on land healing. We've wasted farm business acumen that could have focused on profitability.

We've squandered now a couple of generations by not cultivating a farming can-do spirit. Those best and brightest minds have developed video games, fiber optics, and heart stents. Would we not be a richer culture if at least some of those minds had been encouraged and applauded into farming? Not corporate agribusiness, but farming. My, my, what we might have by now! It's a national disgrace and economic/ecologic travesty that we've denied ourselves these minds because we taught, embraced, and assumed the notion that farming is beneath the dignity and mentality of the best and brightest.

I have a bulging file full of testimonials from farmers who say: "We didn't think it was possible. But when we found out about these systems, we began farming and it's working. We'll be fulltime by next year." Every time I get those letters and see the pictures of bright-eyed beaming children next to their plants and livestock, I get all teary-eyed. I can't think of a more laudable legacy than to leave our culture a legion of best and brightest farmers.

That's certainly better than a legacy of empire building in foreign countries. Or a legacy of bailouts. Or a legacy of government takeover.

I've heard that if your vision can be accomplished in your lifetime, it's too small. Few things excite me as much as meeting sharp young people who want to be farmers. I see it as a reversal of a trend, and a linchpin in the healing of our country. May thousands and thousands of sharp, clever young people join this profitable vocation: lunatic farming. It's noble. It's sacred. It's a great living. It's wonderful scenery. It's a great place to raise kids.

Enjoying this life and encouraging the best and brightest to join in is the sheer ecstasy of being a lunatic farmer.

TAKEAWAY POINTS

1. Do you want your farmer driving a Cadillac?

2. Would you really be happy if your child star became a farmer?

3. Valuing farmers is the cornerstone of environmental protection.

Chapter 19

Relationships

Non-industrial farming is all about cultivating relationships as part of the transparent and open source production and processing lifestyle. Relationships blossom with trust and shrivel with distrust.

Industrial farming cultivates distrust. Ask any farmer who signed up to be a slave for Tyson if he trusts the corporate leadership. I haven't found a single one who thinks company managers are shooting straight. The whole relationship between grower and buyer is based on intrigue. Buyers certainly don't trust the food industry. Ask the average shopper at a supermarket if she really trusts the CEO of a food corporation when he says irradiation is completely harmless.

Industrial food processors consistently abuse their patrons. How else can you explain why they successfully lobbied to criminalize an rBGH-free milk label? How else can you explain why they successfully lobbied to keep genetically engineered foods from being labeled? They don't want any relationship with their customers except one that's mercenary. They don't want knowledge. They want ignorance.

252

That's what is driving the know-your-farmer movement. Thomas Friedman's *THE LEXUS AND THE OLIVE TREE* describes this cultural dichotomy between the appeal of technology balanced by the yearning for roots and heritage. As the agrarian economy gave way to the industrial economy, people fell in love with concrete and rebar. The World War II generation, supposedly the greatest generation on earth, taught the world the meaning of duty. But in the course of fulfilling duty, they forgot something: how to cry.

The hippie mother earth cosmic worshipping tree hugging movement of the 1960s that led to Woodstock and free love was the baby boomer generation yearning for roots. Who would have thought in the 1950s when breast feeding was universally considered barbaric—"haven't we gotten beyond that?"—that by the early 1970s La Leche League and Lamaze would arrive in America and Dads-to-be would actually want to go and see it – "it" being the birth of a child?

The desire to connect with our food is nothing more and nothing less than a primal, soul-level desire for roots. For something more than rebar and concrete. When people know their farmer, they connect viscerally with what is before them on the plate. After all, dining is a fairly intimate experience. Next to the act of marriage, eating is one of the more intimate things we do as humans. We take in this food, right into our bodies, and it becomes us. Flesh. Blood. Being. Mind.

Because it wants no relationship with the eater, industrial food is like prostitution food. No courtship. No romance. No special knowledge and nuances to add delight to the intimate dining experience. Industrial food is like a one night stand. A mercenary relationship. The less knowledge, the better.

Of course, the way things have developed, many people want it this way. "Don't tell me what's in my food," they say. With knowledge comes responsibility. So as long as they're ignorant, they can ingest earth-damaging food and not wrestle with a guilty conscience. Their psyche is free from the horror of knowing their dining aids and abets criminals.

As a lunatic farmer, I want a relationship with my patrons. That creates accountability and encouragement. When Rachel made pound cakes and zucchini bread as a little girl, ladies would buy them and exclaim to her: "Oh, my garden club ladies just raved about your pound cake last week." What do you think that does to the self-image of a child? One of the reasons our young people have such a poor self-image is because we aren't letting them receive adult praise for worthy work accomplished well.

Even as toddlers, our children heard customers tell them how important our family was to their wellbeing. "Our family depends on you for our health and we so appreciate what you all are doing. And our children will depend on you for the next generation..." and on and on it goes. Compare that to the industrial farmers' children who ride in to the grain elevator or the sale barn. Buyers don't care if the farmer succeeds or fails. It's all about the cheapest price possible, and if you don't have it, I'll get it from Argentina or Brazil or Mexico. Doesn't matter.

One of my favorite memories working with our first loyal chef customer, Lisa Joy, was unloading half a beef into cold storage because she didn't have room at the restaurant to store it all. We backed up to this massive cold storage warehouse. My mini-van was sandwiched between two tractor trailers. A forklift operator set a pallet down on the dock and Lisa and I began stacking the boxes of frozen beef on it.

Right next to us was a tractor trailer load of "Frozen Banquet Meals." She and I both had quite a laugh over the juxtaposition. Here were a farmer and a world class chef unloading grass finished beef that would be lovingly and artisanally prepared for discriminating diners. Next to us were 20 tons of boxed, completely processed, nondescript meals. We were lost in a tsunami of industrial food. The whole experience was surreal, like it couldn't be happening. How could we be so insignificant and this industrial counterpart be so astronomical?

Easy. Lots of people didn't care. That's how these things happen. Food is just concrete and rebar. Don't bother me with whether or not we should build here, just applaud the fact that we did it. It's here and it's big. Doesn't matter if it's the wrong thing in the wrong place. It's big, and that should be enough. And it's cheap. Who cares about fertile frogs and four-legged salamanders? They're too small to worry about, too. Lisa and I were just a bug to the industry. Something to be stepped on like a common beetle. Insignificant.

Amazingly, I'm just lunatic enough to love my patrons. To think they are pretty smart. That they are more than billfolds and credit cards. More than pawns to be manipulated by clever slogans and expensive advertisements. More than just mouths to ingest reconstituted processed slop. Lots of slop. The more the better. Add sugar and they'll eat more. I view my patrons as fellow healers. We're on this wonderful pilgrimage to heal health, the earth, our communities, our society. Yes, it's a noble, grand, sacred ministry, and we're moving down this path together. Their children run in and out between my legs as we stand and talk. That's relationship.

I don't have a bevy of Philadelphia attorneys to be a veil of protection between abused customers and me. I'm just out here, vulnerable, doing the best I know how to make healthy partners. Not fatter people. Not pharmaceutically dependent people. I want people to stay out of the hospital. I want my customers to be so healthy they don't go to doctors because all their physical needs are met in this nutrient dense food. I'm not concerned about the quantity they eat; I'm concerned about the quality.

Customers give me gifts. "We want you to stay in business. Here's a little extra," they say after a particularly rough season. Bills. Big green bills. Stuffed into my pocket. Does Dean Foods get that? How about Cargill? Monsanto? Avantis? I don't think so. In fact, lots of people have never heard of these giants that have budgets bigger than half the world's countries. And they like it that way. Anonymity is great for industrial food.

Only a lunatic would want to look customers in the face. Customers are fickle. They sue you. They are hateful, unloving, and vindictive. We had a customer once serve our chicken to some guests who raved about it and asked the hosts where they had gotten such wonderful chicken. The hosts were a little concerned about the character of these folks and decided not to tell them where they got the chicken. The hosts told us later: "We didn't think they would be good customers." Folks, you can't buy insurance with that kind of protection. That's called relationship.

Another kind of relationship farmers have is with their communities, their neighbors. Industrial farms are not neighbor friendly. Certainly a defining difference between heritage-based farms and industrial farms is how they perceive their neighbors and their local community.

If you think back to the social fabric surrounding Pa and Ma Ingalls of *LITTLE HOUSE ON THE PRAIRIE* fame, the camaraderie and deep personal friendships created a seamless social fabric within the agrarian community. There was a depth of interaction and interdependence that is increasingly hard to maintain in a techno-glitzy world.

This was brought home to me very recently by one of our apprentices, who had been a software engineer in the New York banking world—what he affectionately calls his former life. After one especially grueling day here on the farm, he commented that it was still more rewarding because at the end of the day, he could physically see, touch, and handle the result of the day's work.

And he especially appreciated that our team had accomplished it together, with all the head scratching, banter, and sweat that it entailed. In a Dilbert cubicle, many times your team members are miles away. The project is nebulous. It's out there in cyberspace. You can't touch it. And while it took great effort to accomplish, it was only cerebral effort.

The difference is that at the end of the day, the computer technician leaves the office seeking to unwind and to connect with people. Our team finishes the day both physically and mentally tired, having connected with people all day, and is seeking satisfied rest.

I'm reminded of my college debate partner who has built an extremely successful law practice calling me one evening after I'd closed the eggmobile, lamenting: "This isn't fair. Here I am popping Alka-Seltzer and getting ulcers about my clients and the court hearing tomorrow, and you're out putting up the chickens." Yup, that's the sheer ecstasy of being a lunatic farmer.

As a farm becomes more industrialized, interactions with the community inherently diminish. Experts must be called from outside rather than counsel sought from inside. Poultry farms, for example, receive routine visits from the field rep. That's the corporate expert. In many cases, these field reps have never raised a chicken in their lives, but they've been to school and have a credentialed degree.

Field reps come in and tell farmers how to do it. Of course, this is appropriate because most farmers who sign up for a factory house have never raised a chicken either. So the factory chicken farmers, by and large, feel completely overwhelmed by their own industrial paradigm. If they have a problem, they don't call a neighbor; they call the field rep. If anything goes wrong, the farmer can blame the field rep. And of course the field rep blames the dumb farmer. It's a wonderful partnership.

A similar thing occurs in many areas. Our neighborhood mechanic, for example, has a small shop adjacent to his house. A fix-it wizard, he's the kind of perfect neighborhood expertise we need to encourage. But as cars become more and more sophisticated and the technical equipment for reading the computers more expensive, it's harder for him to serve the community. If the computer reading equipment costs $100,000, you need a big shop to pay for it. A one-man backyard business can't afford that kind of equipment.

And while Thomas Friedman, author of *THE WORLD IS FLAT*, seems to relish this movement, I think it cuts at both warp and woof of culture's fabric. In our mad dash to globalism and sophistication, we're not asking, "What about the neighbors?" Culturally, it's as if we respond, "What neighbors?" For all our looking at the other side of the globe, we can't even see the folks who live under our noses.

James Dale Davidson discusses this in his wonderful book *THE SOVEREIGN INDIVIDUAL*, in which he says modern western cultures are re-organizing around value tribes as opposed to blood-related tribes. Historically, people have been organized tribally much longer than we've been organized as nation-states. Chat rooms, twitters, face pages are all part of realigning ourselves with people who think like us rather than people who live near us or are biologically related.

Most of us are much better friends with people who think like us than the yo-yos with whom we must spend Thanksgiving and Christmas. You know, the weird uncle and weird grandma. And weird sibling. A truly happy, cohesive, open family is a rarity. Actually, it always has been. Familial intrigue runs all the way back to Adam and Eve when Adam blamed Eve for his sin.

And Eve, of course, blamed the serpent. Looks like pointing fingers is pretty deeply imbedded in the human psyche.

Here at Polyface we struggle with this neighbor friendly concept because we have had to create our own community. By and large, the old-time farmers in our area fear us and don't want anything to do with us. But when people move into the area, we're quick to collaborate.

In this regard, I have a deep respect for Amish communities. I wouldn't want the legalism, but I deeply appreciate the helpful interaction. I'd love to forget paying insurance to some big outfit because I knew that if my house burned down on Friday, the community would rebuild it on Monday and we'd move in within a few weeks. To me, that's real insurance.

When I hear my neighbors describe the old threshing rings and fall hog killin's, it warms my soul. One of our landlords runs a horse operation and he makes something under 1,000 bales of hay each summer. Do you know what fun it is to go over there with our whole apprentice/intern crew and put those bales in the barn for him? He brings a cooler full of drinks and everyone sweats together until all the hay is in and the barn smells delectable. The comaraderie is palpable.

But most farmers don't do that anymore. They've all gone to round balers to make haymaking a one man operation. The whole focus is to get rid of labor. And so our rural neighborhoods are full of teenagers playing soccer and video games while neighbor farmers drive their tractors by themselves and make round bales. Maybe both people prefer it that way: the farmer left alone and the teen unencumbered by meaningful work, just playing life away. But somehow it doesn't seem like that's as good for the strength of the community.

Industrial chicken farms don't even own their own chickens. An off-farm crew comes in to build the factory house. An off-farm crew installs the equipment. Chickens arrive from the vertical integrator. Their ration is formulated by the integrator: the average farmer doesn't have a clue what's in the feed. An off-farm crew arrives eventually to load up the birds and take them to the abattoir. The farmer doesn't know the names of any of these people nor where the chickens are going.

Meanwhile, the stench of this factory house has ruined every Sunday picnic downwind for two miles. And when neighbors complain, farmers retort sanctimoniously: "Let them starve and then maybe they'll appreciate me." It's as if the industrial model is so focused on export sales nobody even stops to consider what's happening right in the community.

How else can toxic dump sites and corporate pollution be justified? Right here in our county the South River still has astronomical levels of mercury due to DuPont's dumping there half a century ago. Nobody knows how long it will take for all that mercury to dissipate and for the fish to be healthy and edible again. I agree with Wendell Berry that planetary stewardship starts in our backyard. How in the world can we expect to nurture the world if we don't first nurture our own community?

Nurturing means we keep our backyards healthy. It means not turning them into toxic waste sites. It means maintaining aesthetic and aromatic beauty. And if we can't do that, we need to stop doing anything until we figure out how to do it.

Nurturing means hiring people in the community. Using the local labor force, even if you have to pay a little extra, will always yield more stability and security than bringing labor in from outside.

Nurturing means staying within the carrying capacity of the region. Our county had a brush with disaster a couple of years ago when local government officials began courting a Toyota plant to locate here. Of course, the plum was jobs. But the project was going to take three farms by eminent domain and increase water usage by 50 percent of the entire county's current usage. Can we never learn the lessons of carrying capacity?

I'm thrilled to know that my kind of farming doesn't pollute the groundwater. It doesn't require a cleanup effort down the road. And anyone in the country would love to live right next to Polyface.

Healthy farms cultivate on-farm people relationships more than an industrial farm. While industrial farms pride themselves in fewer people, at Polyface we encourage more people. Immediate family, extended family, interns, apprentices, subcontractors and other team members create a multi-aged, multi-talented, multi-interest group. I enjoy the human relationships on our farm much more than the relationship with a tractor. A tractor doesn't cry with you. A tractor doesn't encourage you. A tractor doesn't give you a hug. People are generally more personable than tractors. Rather than figuring out how to have fewer people on the farm, lunatics try to increase people on the farm.

Certainly having more people creates its challenges. People have feelings and tractors don't. The relational finesse required to maintain a cohesive team pays for itself in heart level sharing and caring. While the U.S. domestic agricultural policy touts fewer farmers as a benchmark of success, I think it's a travesty.

If you visit the average farm at 1 p.m., you can honk your horn, but nobody is around. Only whirring fans or feed augers running. The fields are devoid of people. An occasional crop duster makes an appearance before disappearing over the horizon.

My favorite scene in any movie is the one in *SHENANDOAH* when Jimmy Stewart asks his sons which ones want to join the army. He's in the barnyard and begins calling them by name. Two poke their heads out of the hay mow. One comes out of a door carrying a bucket of milk. Another comes in from the chicken yard carrying a bucket of eggs. One by one the sons enter the barnyard from their respective chores. It is one of the most gripping, warm scenes I think ever created in a movie. That's all you need to know about my agenda.

While most farmers are trying to get rid of people, I'm trying to bring more people in. Most farmers don't even want their children to be involved in farming because the work is long and hard and doesn't pay. Most farmers encourage their children to go do something else with their lives. And most children oblige. After all, most children who grow up in a home where the parents complain about their vocations don't end up pursuing those vocations. They do something else, understandably.

From day one, I have talked up farming. I think it's the greatest life in the world. I want my kids to love farming. I want my grandkids to love farming. I want other people's kids to love farming. I think the greatest resource loss in America is not soil or virgin forests. It's the loss of people on farms. People who have been replaced by machines. And now machines that aren't even run by people, but by global positioning satellites. Driverless tractors. Imagine that.

The capacity to love and observe is much higher in humans than in machines. I think farms populated by loving stewards is a beautiful sight. On our farm, when you show up at noon and honk your horn, I want a dozen people to come out of their respective work stations, big happy smiles across their faces, ready to welcome you and show you their latest endeavor. The older I get, the more I appreciate youthful energy. Surrounding myself with enthusiastic youthful energy, leveraged meaningfully by my experience, is definitely better than growing old lonely and grumpy.

Some people say Polyface has an army of workers. Do you know what we can get done in a day with a focused army? We unleash all that healing energy on this landscape and things change. It's an awesome thing. Most farmers don't like people. That's why they are farmers. They want to go over the hill, commune with their tractor, and let the world go away. Yes, I like being alone once in awhile too. But it sure is nice having a couple extra sets of hands for most projects. Two plus two often equals five.

Jobs that would be drudgery turn into a party when several people are working together. Joking, banter, and good-natured conversation make chores a time of enjoyment. What would otherwise be a long row of beans to pick becomes a shindig when several people are picking together. An otherwise daunting task can be done in an hour when several hands join together.

I'd like to see everyone who ever wanted to grow something out on a farm doing it. Why should all this pent up yearning be sequestered in some sunless, lifeless office at the end of an expressway? Unleash that desire. We should have hundreds and thousands of land lovers dancing across the countryside, dancing with earthworms, dancing in eggmobiles, dancing with pigaerators, dancing with vegetables.

I was on a farm in Illinois, a huge organic grain farm. The owner said back in 1900 that farm supported ten families. Today it supports one. So few people are on the land now in his area that three school divisions had to consolidate just to fill one school. Children ride more than an hour one way to attend school. Think of the community, the wisdom, the laughter that left those fields as those diversified smaller farms gave way to industrial consolidated farms.

I was on a ranch in Colorado. The manager took me to the school house that served the families in the community. Now the community is gone and the one ranch covers the entire 100,000 acres. One or two families live on the ranch. That's it. Lack of proper grazing management gradually destroyed the productive native grasslands. As the land moved toward infertility, it could no longer support families. The solar metabolism shut down and the families left. That is not a good thing.

Many times the reduction in farmers and ranchers is not due to mechanization, but rather due to fertility depletion, especially in brittle environments. I'm certainly not saying everyone should be a farmer, but in order to be farmed and ranched well, and to be indicative of healthy landscapes, rural America should be home to thousands of additional farmers.

Beyond the human relationship elements, though, lunatic farmers embrace intricate relationships between animals and plants. Multi-speciation is all about massaging relationships. Crop rotations. Diversity is the crux of building relationships. Industrial farms hate diversity.

According to John Ikerd, retired agriculture economics professor and sustainable agriculture prophet, the underpinnings of industrialization are specialization, simplification, routinization, and mechanization. But biological systems are not like that. Biological systems are not industrial systems.

Biological systems are diverse, not specialized. They are complex, not simplified. They are dynamic, not routinized. And they are spontaneous, not mechanized. Just because something works in a factory on a machine does not mean it works in the biological community. To impose industrial principles on the living world is to deny everything we know about biology. And yet that is the premise of industrial farms.

Farmers, in fact, are known by the commodity they produce. Farmers introduce themselves as dairymen or vegetable growers or orchardists. Nothing could be further from nature's template. Living systems have intricate relationships, choreographed carefully with checks and balances to preserve diversity within the whole. In the industrial mindset, no field is too big, no confinement animal factory too crowded, and no processing plant too gargantuan. Nature sends disease to check inappropriate growth.

I wrestle even identifying myself as a pastured livestock producer because that sounds confining. I prefer to just say I'm a land healer. Why would anyone want to be called a cattleman or a vegetable grower? How about a caretaker for a piece of God's creation? How does that sound? We may find that the animal or plant we've selected is not a good fit for our area or our temperament. What then? When we locate an orchard in a frost pocket, do we repent and relocate it? No, we use energy intensive technology to keep the frost away. If we had not been so brash and arrogant as to think we could make an orchard wherever we wanted it, maybe we wouldn't have located it in a frost pocket.

All of these relationships, though, in the modern farmer's mind, add unnecessary complexity to the operation. Farmers like it simple. But nature isn't simple. A natural farm must inherently be complex. You can't have a natural farm with as simplistic a production model as an industrial farm. Although the relationships sound difficult, with time they get easier, just like anything. Remember, anything worth doing is worth doing poorly first.

As you move past the simple types of critters, though, relationships become even more subtle. For example, the relationship between a hen and her nest. I confess that I wouldn't fall out with anybody over this, but at Polyface we don't use roll away nest boxes for laying chickens. Of course, in the industry the birds are in wire mesh cages and don't even have nests. They just squat and drop the egg on the wire. The wire floor is slanted so the egg rolls out under one edge and onto a conveyer belt that takes the eggs to a processing room. It's all quite efficient.

But nobody asks the hen about her relationship with anything. Her environment. Her egg. Her cellmates. The sun. Her feet. Her debeaked beak. I'm a bit chagrined, however, at the number of heritage-based or pasture-based egg operations that are using roll away nests. In these, the hen enters a slanted-floor nest box and the egg rolls out into a covered holding area. These nest boxes have a plastic astro-turf kind of bottom that allows the egg to move away to the adjacent holding area.

Have you ever seen a chicken on a nest? First, given the chance, she never nests on a slant. She always finds a level place. But beyond that, she spends a fair amount of time getting situated. She kind of hunkers and wiggles and picks up pieces of straw or hay and places them around just so. She is giving birth, and that takes time and attention to detail. It's not just a mechanical process. She has to get things ready. You can't just go lay an egg half-cocked (pun fully intended).

You've got to place your luggage just so. The temperature must be right. You need some privacy, so you've got to scooch around just right. There, pick up that piece of straw and lay it over there. No, preen just a bit. Pick that pesky bug off the side of the nest. Yum. Move that other piece of straw. Now we're getting settled. Finally, and only after all is well, the egg pops out and the hen leaves the nest, that steaming, moist orb lying magnificently on the perfectly formed, level nest.

I don't know what building that nest has to do with the essence of egg. I'm not sure you could measure it with any empirical test. But when I think of relationships, I can't help but think of that nesting relationship for the hen. Denying her the chance to build that nest, to move things around, to get settled on the level...well, it just seems like she ought to be able to do all that. Ultimately, this level of relationship-massage will show up in the relationship I have with my patron. It always does come full circle. Everything relates to everything.

Finally, as we dig a little deeper, I need to foster my relationship with the microbes. Those unseen critters busily working in me, around me, under me, and over me every day of every year of every decade. This busy bevy of bugs, eating, killing, desiccating, mating, jousting, going to school, filling out IRS paperwork. My goodness, it just dawned on my why these guys get so much done. No wonder they work for free 24/7. They don't have an IRS. Now it all becomes crystal clear.

Anyway, these microbes have relationships with each other, with me, with the plants and with the animals. Keeping the good ones thriving, happy, and well fed, happily housed, is key to keeping the few bad ones in check.

This is one of the reasons children should be in the garden. They need to get in the dirt, get scratches and splinters, a few blisters. Not only does this exercise their immune system, but it also builds a relationship with this microbial world. When we nuke this entire world with anti-microbials, we set ourselves up for bad guy invasion.

Beyond that, though, children need to learn that they are not the center of the universe. When all you do is create your own world out of computer fantasy, you enter life pretty jaundiced about reality. In a video game, if your car wrecks, you wait a couple of seconds and get a new one. If you get killed, the game gives you a new person. You create your own reality. That's dangerous, because the world is bigger than we are.

Disease happens. Drought happens. Frost happens. Hail happens. The lettuce bolts. The beetles find the potatoes. And children need to know that they can't just tap a button on a console and have the world bow to their whim. Gardening builds relationships with microbes, with reality, and with humility. Our world needs a lot more humility and a little less hubris. Gardening does that for kids.

Dirt is great on kids. It will wash off. What do we teach our children when we call soil yucky? Parents should do everything possible to encourage their children to build relationships with their ecological umbilical. Eat a little soil. It's good for you. Get those microbes in there. Exercise that immune system. Wiggle on down in that nest. Move some stuff around. Build that relationship.

Industrial farms want to annihilate microbes. They want sterility. And they think they can substitute a relationship with microbes by cultivating a relationship with pharmaceuticals. I challenge anyone. Take a handful of good compost, full of microbes. Bury your nose in it and inhale deeply. Now take a handful of any drug, any pesticide, any chemical fertilizer. Bury your nose in it. Inhale deeply. Which relationship would you rather have?

My deepest prayer is that I will faithfully cultivate all the relationships I'm supposed to—both known and unknown. Life is grander than any of us can imagine. To bully around as if relationships are not delicate, marvelous things is to miss the mystery and splendor of the living world. Appreciating and loving that world is the sheer ecstasy of being a lunatic farmer.

TAKEAWAY POINTS

1. Healthy farming needs more farmers, not fewer.

2. Biological beings cultivate more meaningful relationships than machines.

3. Chickens like to build a nest.

Chapter 20

Direct Marketing

Modern American farming is part of an industrial model that feeds raw commodities into processing plants that cook, break, package and reformulate to supply wholesale products to retailers and resellers. The entire system recognizes only commodity type and does not differentiate based on nutritional quality or ecological stewardship. It is a completely industrial system.

But it hasn't always been this way. I'd like to quote from the 1942 agriculture college textbook, *FARM MANAGEMENT AND MARKETING*:

> *As contrasted with city industry and the merchandising of factory-made products, farming is done by a large number of small units in which the individual farmer is both a laborer and the owner of capital invested in the business in which he is working.*
>
> *Farming is not adapted to large-scale operations because of the following reasons.*

268

1. *Farming is concerned with plants and animals that live, grow and die.*
2. *Farming is greatly dependent upon climate and weather conditions.*
3. *Farming requires the making of many quick decisions by the individual worker.*
4. *Farming requires enough personal interest on the part of the individual worker so that he will assume major responsibilities and work harder or longer hours on some days than on others.*

Can you imagine this being the official collegiate position of the USDA and agriculture college professors? We've come a long way, baby. Now you see why I don't look at anything published by the USDA after 1950. It all went to pot after that.

In those days, as America was urbanizing but before food industrialization many, if not most, farmers direct marketed to end users. In 1950 D. Howard Doane wrote *VERTICAL FARM DIVERSIFICATION: ADDED INCOME FROM GRADING, PROCESSING, DIRECT SELLING.* The 1950 publication date skews the real timeline of Doane's revolutionary thinking and writing, which took place from the late 1930s and through the 1940s. Here is a quotation from the book:

Those who are only producers of raw farm products are in an unbalanced, hazardous position. During most periods they are at the mercy of purchasers. The products respond most rapidly and drastically to price fluctuations.

Doane continues:

In a recent issue of the PROGRESSIVE FARMER, Dean Chapman quotes George H. Stevenson as follows:

The tendency of civilization is to make of the farmer a producer of raw material solely, with the manufacturing and distribution entirely in the hands of the highly organized, but not necessarily efficient, urban centers. No industry or nation can long survive solely on a basis of production of raw materials, leaving in other hands the marketing of the material in its raw state, as well as the manufacturing and final distribution to the ultimate consumer. It is the history of both nations and industries following this course, that the producer of the raw materials becomes steadily poorer, while the distributor and manufacturer become richer and more powerful.

Will you indulge me one more from Doane? Here is a man who ran in top agricultural circles, at the federal level, before 1910. What I'm passing on to you, in these little vignettes, is the culmination of a lifetime of thinking from a man who was at the forefront of American agriculture, who had the ear of presidents and was personal friends with secretaries of agriculture, from 1900 to 1950. I think it behooves us to appreciate what a man like this has to say toward the end of his life:

...it is clear why manufacturers press down the prices of the raw products they purchase. Who, then, must make the concessions? The answer is those who are least able to resist pressure and who can take reductions and still stay in business. There is but one group that regularly meets these specifications— farmers. They are the only ones in the long chain of those who handle raw products who can take reductions and not go out of business. The farmer can do it because many of his cost-of-production items are hidden and can be minimized and put off. He can sell the fertility of his soil without restoration for long periods. He can call upon his family for labor without cash payments. Repairs and replacements of buildings can be delayed. Depreciation is seldom, if ever, counted as part of cost. When the transporter, warehouseman, or processor figures costs, all items, actual and calculated, are a part of the formula which makes up the price he sets.

Isn't that great stuff? He goes on in the book to explain how the whole system is stacked against the farmer and advocates farmers becoming their own middlemen. We have a saying: "The middleman makes all the profits." Hey, if that's the case, I want to be the middleman. Fortunately, back in his day, the food police had not swelled their ranks and written as many regulations as they have today, and many of the illustrations Doane uses throughout the book to show how farmers can direct market are now illegal.

He talks about home canning and sending boxes of home-canned pickles and meats through the mail to urban customers. That's illegal today. He talks about curing hams at home and selling them to restaurants in town. That's illegal today. My father-in-law remembers how cured hams kept their farm in business during the 1930s and 1940s, even up into the 1950s. He carried many a ham on his shoulder to restaurants in Staunton. These hams were from hogs his family raised on the farm, butchered on the farm, and cured in a smokehouse out behind the farmhouse. All illegal now.

In this walk down memory lane, I'd like to go back one more notch, to one of my favorite little books, *MAKING THE FARM PAY*, by C. C. Bowsfield, written in 1913:

> *The average land owner, or the old-fashioned farmer, as he is sometimes referred to, has a great deal of practical knowledge, and yet is deficient in some of the most salient requirements. He may know how to produce a good crop and not know how to sell it to the best advantage. No citizen surpasses him in the skill and industry with which he performs his labor, but in many cases his time is frittered away with the least profitable of products, while he overlooks opportunitites to meet a constant market demand for articles which return large profits.*
>
> *Worse than this, he follows a method which turns agricultural work into drudgery, and his sons and daughters forsake the farm home as soon as they are old enough to assert a little independence. At this point the greatest failures are to be recorded. A situation has*

developed as a result of these existing conditions in the country which is a serious menace to American society. The farmers are deprived of the earnest, intelligent help which naturally belongs to them, rural society loses one of its best elements, the cities are overcrowded, and all parties at interest are losers. The nation itself is injured.

Farm life need not be more irksome than clerking or running a typewriter. It ought to be made much more attractive and it can also be vastly more profitable than it is. Better homes and more social enjoyment, with greater contentment and happiness, will come to dwellers in the country when they grasp the eternal truth that they have the noblest vocation on earth and one that may be made to yield an income fully as large as that of the average city business man...

Raise a first-class article, whether grain, vegetables, chickens or pigs, and there will be no difficulty in finding people who want your product if you will but let them know what you have and what you sell it for.

I have often seen men going from store to store with a tin bucket and an old rag sticking out under the cover asking the merchants if they wanted butter, and at every place they would be told that it was not wanted, when in fact those very merchants were getting print butter all the way from Wisconsin or Iowa. They knew the character of the butter in the tin buckets and did not want that sort. As with butter, so it is with all products of the farm. It is quality that makes the article sell.

Conditions are right for money-making by the agricultural class. It simply remains for the farmers themselves to develop methods of selling by which they can take advantage of the improved markets. The rapid growth of cities, and the sharp demand for all kinds of produce are substantial evidence of this improvement....

It ought to be the aim of every farmer to accomplish these definite results:

Increase profits by enlarging production at a fixed expense.

Diversify crops and all other profits so as to distribute labor evenly throughout the year.

Secure a regular income at all seasons by supplying customers with poultry and dairy products, vegetables, beef, pork, etc.

Shorten the work-day to ten hours, provide a comfortable home, improve the appearance of the premises and try to make life enjoyable.

Let the young people have a little money from the production of fruit, flowers, vegetables and experimental crops.

Teach them to plan work for themselves and to love the country.

Isn't that great? I do not apologize for such a long excerpt. This book, now a century old and out of print, is as current today as it was then. As a tribute to the lunatics on whose shoulders I proudly stand, I wanted to introduce you to a smidgen of this rich lunatic heritage. You can see why I love these guys. Those of us on the lunatic fringe have been preaching the truth for a long time. My hope is that likeminded people will proudly adopt the lunatic mantra and keep preaching. My goal is to create a whole generation of lunatics.

The point of all this is that to promote farm direct marketing, you have to be a lunatic. If the modern industrial food empires had their way, visitors and customers would be excluded from farms. They view anybody coming onto the farm as threatening the world's food supply. Mark my words, the days are fast approaching when you will see legislation to prohibit all farm access to anyone but the farmer. This is called sophisticated, science-based farming.

Any farmer who promotes direct marketing, having people crawling all over the farm, and developing a farm sales plan, logo, and sale tax identification number is definitely a lunatic in today's world. Remember, for all the interest in local food, it only represents about 1.9 percent of food sales. All the farmers' markets, Community Supported Agriculture (CSA) farms, farmstands, direct farm distribution put together scarcely reach 2 percent of food sales in America. It's still very small.

To be fair, the downside of direct marketing is that production is limited to marketing. One of the advantages of regular commodity farms is that the buying universe is big enough to absorb any single entrant's production into the system. If I added 1,000 beeves to the system tomorrow, it would handle it without a blip. If you're growing cabbage for Del Monte, a few tractor trailer loads more or less won't even create a hiccup in the market.

But in direct marketing, if I produce one more dozen eggs than I can sell, its value just went to zero. If I produce one chicken beyond my market, its value is zero. That's the hard truth about direct marketing. If you figure out how to really produce, you have to wait for the market to catch up. And that can often be a long wait, unless you're a crackerjack marketer. Which brings me to why I promote direct marketing.

Today's conventional farm has an income stream from one thing: production. But raw production is subject to the four horsemen of the Apocalypse: weather, price, disease, and pestilence. These are the topics of conversation whenever farmers gather. And for the most part, they are things farmers can't do much about. The bottom line is that farmers spend most of their lives complaining about things completely beyond their control.

That farmers would dwell on these things is understandable, though, since all their income is subject to these vagaries. Income derived from production is vulnerable to these unknowns, and that is why I encourage farmers to spread their income to less risky parts of the food dollar.

In very general terms, the food dollar gets split four ways: production, processing, marketing, distribution. While these are not always clearcut delineations, they represent the big picture fairly well. Each of these sectors accounts for around 25 percent of the retail dollar. Production is the one most susceptible to those whimsical elements we talked about a moment ago. If grasshoppers come, they will destroy a field of corn way before the combine that harvests the corn, the elevator that stores the corn, the fans that dry the corn, the equipment that makes cornflakes, the trucks that carry cornflake boxes to the supermarket, the advertising agency that promotes the cornflakes, and the supermarket that sells them.

The lion's share of the vulnerability in the food system is loaded on the front end at the point of production. That's the most risky part of the chain. Once the raw material is in the food chain, it's fairly secure. Yes, things can happen, but not with the regularity of weather or disease events out in the field. Think about the legs of a stool. A one-legged stool can tip over pretty easily. And that's what most farms have—a one-legged income stream.

Think of the other legs as processing, marketing, and distribution. Farm income becomes more insulated from nature's whims when more dollars derive from these other sectors. The farm actually becomes much more stable as a business by diversifying its income streams to less vulnerable food economy sectors. Most farmers, of course, do not want to be processors or marketers or distributors. That leaves more room for the lunatics who do.

When a farm derives income from these other three sectors that account for 75 percent of the food dollar, it builds additional legs on its income stool. A four-legged stool is much more balanced than a single legged one. And those other three legs are not as vulnerable to weather, price, pestilence and disease.

To be fair, if a farm is located 200 miles away from a Coke machine, this becomes more difficult. And I'm quick to admit that I don't have all the answers. Our farm is on a dirt road where the only time you have to lock your car is in August to keep the neighbors from putting runaway zucchini squash in it. Polyface is not on a main drag. We're nestled in here against a mountain on the noncommercial end of town.

One of my favorite sayings is from Ralph Waldo Emerson: "I trust a great deal to common fame, as we all must. If a man has good corn, or wood, or boards, or pigs to sell, or can make better chairs or knives, crucibles or church organs, than anybody else, you will find a broad, hard-beaten road to his house, though it be in the woods." Notice the accent here is on *high quality*. That precedes, drives, pushes everything else.

That's the problem with organic certification as a marketing tool. Now that the government owns the term and defines the protocol, the same kind of collusion toward shortcuts that defined corporate/government agendas now defines organics. For the life of me I can't figure out why people who fought the USDA for decades because it pooh-poohed everything nature-respecting suddenly decided to turn over to the USDA the reins of organics. That's called intellectual schizophrenia.

More skill diversity is necessary if a farmer wants to enjoy retail dollars, more stable income, and more business balance. Since most farmers do not want to augment their skills with anyone else, their farm business is self-limited. It distresses me when I hear farmers say: "This farm is just Matilda and me and nobody else. And I don't want anybody else." Well that might be fine when you and Matilda are in your 30s. But what about when you hit 60? This is the harsh conundrum facing more than half of America's farmers today.

I have a stack of letters from elderly farmers looking for young people to inherit their farms. They don't want to give the farm to their children because the children will just sell it and cash it out. These elderly farmers love the land, but they never built a business. They built a job for their lifetime, but not a multi-generational business. A farm that diversifies into these other income streams becomes more successional. A farmer surrounded by a team of multi-aged multi-talented people has a much better chance of finding replacements.

An actual farm business becomes a going concern. Going concerns have a much greater chance of surviving the initial entrepreneur than one that kind of fizzles out with the energy of the aged founder. That's exactly what is happening to thousands and thousands of farms across America. It's a time of unprecedented land ownership transferring out of the WWII generation. And it's a time of unprecedented opportunity for young people wanting in.

And that brings me to the next big reason for direct marketing. In addition to spreading income to less risky portions, the very nature of direct marketing creates big enough challenges to attract the best and the brightest. Designing logos, creating sales pitches, figuring out marketing angles and distribution efficiencies—these challenge the best and brightest.

Interacting with customers, writing newsletters, creating brochures, hiring delivery drivers, and boxing with bureaucrats require many different skills. They are the kinds of skills that attract young people to any business. Dealing with these things is a constant challenge because they change from day to day. Once you learn how to plant carrots, the protocol is fairly routine. The seeds always look the same. The planting bed looks the same. The carrots look the same. The weeds around the carrots look the same.

But when you're dealing with people, curve balls are normal. On our farm, my special joy is watching our interns and apprentices interact with customers. The discussion often moves from politics to predators and then to recipes, nutrition, education, retirement plans and ends with earthworms. All this in a 15-minute conversation. Seldom are carrots that cerebrally stimulating.

The well rounded farm, then, requires these diverse talents as a matter of course. For the sake of clarity, here is my list of minimal requirements for a direct marketing farm. Think of it as a pie. In order to have a whole pie, without a slice missing, here are the six slices you must have.

1. PRODUCTION Obviously, you can't sell food you don't have, so somebody has to produce it. In order to be people friendly, this production must be aesthetically and aromatically romantic to the senses. People don't want to come to a stinky farm. The production should be beyond organic. If it's livestock, it needs to be pasture based. Customers don't want to come and see a feedlot—even an organic one.

If you travel by the organic confinement dairies out west, you can't tell a bit of difference between those and the conventional ones next door. They look the same and smell the same. Direct marketing farms really have to walk the talk because customers certify the veracity of what's going on. Customer certification insures the highest level of integrity.

Production must be transparent and open. Here at Polyface, we have a 24/7/365 open door policy. Anyone is welcome anytime to see anything anywhere. A couple of times people have misinterpreted what they've seen. But once we explain, they're usually okay. Yes, it's risky. Insurance companies hate it. But a farm without open doors is untrustworthy, period.

2. **PROCESSING** With today's dearth of domestic culinary skills, for the foreseeable future somebody is going to have to prepare food close to where it is eaten. Here's my totally sexist statement: thirty years ago every woman knew how to cut up a chicken. None of our customers (99.9 percent of whom are women) ever asked how to cut up a chicken. Today, many of them don't know that a chicken has bones. Honestly, if it's not boneless skinless, they don't know what to do with it. It's sad.

But what that means is that somebody has to get it killed, churned, cooked, aged, diced, pureed or something. I'd like to see commercial kitchens on thousands and thousands of farms across America. Most farmers love production but hate processing. It's often tedious, oven-type culinary work. But it's the key to broadening your market base. Every time you add a processing step, you multiply your potential customers. Very few people any more will buy half a beef at a time.

When I began direct marketing back at the end of the 1970s, our whole farm business was based on beef halves and quarters. To sell that volume at a time requires the customer to have a freezer and some disposable cash. Today, very few people have enough freezer capacity and even fewer have $500 at any one time. By breaking it down and selling it in pieces, even though the price is 20 percent more, way more customers will buy. And of course that offers the opportunity to cherry pick. If all they want is filet mignon, that's all the customer needs to buy. Customers love choice, and processing offers more of that to them. We still sell halves and wholes to people who want quantity discounts, but far more is sold in pieces.

3. **ACCOUNTING** Somebody has to balance the checkbook and watch the money. Here again, most farmers hate doing this. I'm blessed with a wife who will spend a day looking for a penny. Teresa is meticulous and accurate, two characteristics you want from someone watching the money.

The accountant needs the authority to authorize expenditures. If you're running a business, you need to know what's in the bank. An accurate understanding of the money flow is essential, and if nobody in your farm business wants to do it, you need to find someone to do it. Otherwise this essential piece of the pie is missing, and you don't have a whole pie.

4. **MARKETING** If your farm team doesn't have a gregarious storyteller schmoozer, you'll never sell successfully. I'll give you one guess who it is in my family. Anyway, marketing is never easy. It's hard work. Challenging and satisfying to those of us who like it, but torture for those who don't.

Of course, marketing is a lot easier when you have a noble story and sacred mission. Healing the earth and healing people is fun to sell. Marketers must be extremely optimistic. Marketers are people who call them go lights instead of stoplights. Everyone will not buy from you, no matter how good you are. I don't dwell on the people who turned down Polyface product. I can't imagine why they would, but it's a big world out there and everyone has his reasons.

Farmers often don't make good marketers because they are so emotionally tied to what they are selling that they take rejection personally. "Don't you know how I sweated over these beets? I missed supper twice to keep the moles out of them. Had to tote water in teacups to keep the seeds moist. Blah, blah blah..." That fear of rejection is why most farmers don't want to market.

If you don't have a marketer in your outfit, go find one. Maybe a defunct Amway distributor who listened to all the marketing tapes on how to sell a hat rack to a moose. Or ice cubes to an Eskimo. If you think you can just produce it and they will come, forget it. They won't. Read Zig Ziglar and Dale Carnegie. Then go do it.

5. **DISTRIBUTION** The food needs to rendezvous with the customer somehow. Whether this is something you do personally, hire out, or collaborate with someone already running a network doesn't matter. But realize that when it comes to distribution, it is cheaper by the million and the ton because the big cost is the labor.

At Polyface, we've enjoyed pioneering what we call the Metropolitan Buying Club (MBC), which is just a glorified way of saying urban drop points. Demanding everyone to come out into the country every time they want a dozen eggs is neither realistic nor sustainable. Somehow we need to duplicate the efficiencies of industrial distribution, but on a local scale. That requires networking and using e-communication to create real time diversified inventory food commerce.

The conundrum facing all of us in this movement has been this: how do you take a story-based open-sourced Eastern-holistic integrity food item and preserve that in a Western compartmentalized opaque dishonest supermarket setting? It's hard. I tried to get into our local Kroger last year and couldn't get past first base. I just gave up. I quit trying to dance with the industrial system. Let's just make our own system.

But people need lots of choice and they need convenient service. We deliver to these drop points within four hours of our farm, and collaborate with other farmers to add volume and variety. Distribution permutations are many, but at the end of the day, the customer can't be required to jump a million hurdles to get local food.

This is my problem with most farmers' markets. They are destination places, often held on Saturdays, and almost never at night when people are out and about shopping or attending little league games. If we are ever going to penetrate the market, we need to have more customer friendly interfaces. Otherwise, we will only sell to the choir. We need to sell beyond that. Distribution is probably a bigger hurdle in the local food movement than production.

6. **CUSTOMERS** Without customers, you can't have a direct market farm. The closer the farm is to a metropolitan area, the easier finding customers becomes. But more importantly, a direct marketing farm needs the right kind of customers. Prima donnas aren't the answer.

We need customers who love their kitchens. We need customers who enjoy trying new things, who will try to use the entire vegetable or the whole chicken. The less processed the food, the better it is for local farmers.

We need customers who put a high priority on food and who want farmers to enjoy a white collar salary. A direct market farm doesn't need a bunch of whiners and complainers. That said, I've been told that if 10 percent of your customers aren't complaining about price, you're probably not high enough. Direct farm marketers need customers who show up at rendezvous points on time, who chat you up to co-workers and neighbors, and who forgive the occasional mess up. One thing's for sure, any farm that heads down this path is going to make a mistake now and then.

Part of customer relations is being quick to apologize and fess up about a mistake. But at the same time, customers who scream and rant and rave should go to Wal-Mart. We pull them out of our customer box. Gone. We get rid of them. I've been in airports a few times and thought: "I wish that agent had the authority to just tell that guy to take a hike. I mean, tell him your airline won't serve him anymore. He's a pain customer and not worth the hassle. Good riddance, buddy." I don't think the customer is always right. This is a partnership, not a dictatorship.

In my view, those are the six essentials of a whole farm direct marketing pie. Someday I hope to write my relationship marketing book, so I'll not put all the secrets in here. After all, I need you to buy another book.

A lot of people quickly snub this direct marketing idea: "Everybody can't do it." Yes, that's right. Everybody can't, and everybody won't. But if everyone who could would, the food system would so radically change that we can't imagine what kind of opportunities would pop out the other end. In order for something to be a great idea, you don't have to figure out how everyone can participate. If it's good for one, that's all that matters.

If we would be as quick to realize the strength and validity of direct marketing as we are to shoot it down because it doesn't have universal application, we'd be realigning the food system so fast the Cargill board of directors would be heading for Pakistan. And the Monsanto board would be right behind them. They'd all be bailing out—unless of course, our government decided they were too big to fail. What happens when the people dessert obsolete multi-nationals?

Probably, lots of those board members would become honest backyard compost-grown tomato producers. Wouldn't that be a neat development? Too big to fail means too big to be held accountable when the marketplace speaks. When enough people realize how wonderful building relationships with direct farm marketers can be, the current power brokers in the industrial food system will either develop integrity, or will tumble into bankruptcy. Either one would suit me fine.

Direct marketing is not for the timid or paranoid. If you walk around every day looking over your shoulder expecting someone to sue you, probably direct marketing is not for you. The closer producer and patron are, the greater the respect, accountability, and loyalty. A short food chain is much stronger than a long one.

People have told me I'm not a very good farmer, but that Polyface success has been because we're good marketers. I don't think that's true, but even if it is, what it shows is that even a mediocre farmer can still be highly successful by enjoying a four-legged stool. And that's the sheer ecstasy of being a lunatic farmer.

TAKEAWAY POINTS

1. The USDA has not always endorsed industrial thinking.

2. Direct marketing has always been for lunatics.

3. The six components of a local direct marketing farm business are production, processing, accounting, marketing, distribution, and customers.

Chapter 21

Localized Economy

Even more than farming without chemicals, the idea of a localized economy whips the industrial foodists into a frenetic frenzy. Localization is archaic in modern America; globalization is the new hip. I'm well aware that pro-localization is not just out of place in agriculture. It's out of place everywhere. But since food exports have and continue to be the linchpin of America's balance of trade agenda, localizing food comes as close to striking at the heart of modern America's paradigm as anything.

Probably the most readable and erudite treatise on this topic is Michael Shuman's book *THE SMALL-MART REVOLUTION: HOW LOCAL BUSINESSES ARE BEATING THE GLOBAL COMPETITION*. In this well researched book, Shuman introduces the term TINA—There Is No Alternative. Even people who intuitively question the wisdom of global everything often eventually shrug their shoulders and retreat into TINA. The die seems cast, and here we are on a nonreturnable journey into the belly of globalization for everything.

He says globalists believe that "community is just another obstacle to progress." Those of us who defend localization are

284

likened to Don Quixote, tilting at windmills. Michael introduces his term, LOIS—Locally Owned and Import-Substituting, as a viable alternative and does a powerful job defending and promoting it. My point here is that excellent business minds and models already exist to prove the efficacy of LOIS. I'm not tilting at windmills.

The average morsel of food in America travels 1,500 miles. The average calorie of food requires 15 calories to get on the plate. Four of those are in transportation. The food system is increasingly opaque, centralized, and distance-oriented. In both Canada and the U.S., at most only 5 percent of food eaten in any area is actually grown in that area. What that means is that trucks are passing in the night. The produce of one area is being trucked to far off places, while produce from far off places is being trucked into the local supermarkets. And this is efficient? And reasonable?

Just yesterday I received a letter from a 14-year old student in New York asking why a tomato from Peru is less expensive than the locally grown one. A lot of that has to do with slotting fees, insurance requirements, and corporate welfare. The whole system is tilted against localized production and consumption. Because of foreign trade sweetheart concessions and subsidies, it's cheaper to print a book in China and ship it to the U.S. than it is to print it and ship it just a hundred miles domestically.

These are systemic problems. It does make you wonder how far our country has moved from its moorings. The founders envisioned a federal government financed solely by excise taxes, or import tariffs. Today, not only do imports pay nothing, but we finance the federal government on the back of the domestic economy. What an interesting scenario to contemplate: no federal domestic taxation and the federal government completely dependent on import tariffs.

I actually don't know if I'm a protectionist or not. But such a scenario would either greatly shrink the federal government— which would be a great thing in my opinion—or it would create a perpetually blossoming domestic manufacturing sector. That probably wouldn't be bad either. Oh, the what if's of history. Most Americans today have no clue that the country performed very well for a century and a half without an IRS. Fancy that.

During that time, the butcher, baker and candlestick maker were securely imbedded in the village economy. This proximity encouraged transparency, which is the linchpin of accountability, which is the cornerstone of integrity. As the scale of each of these enterprises increased during the industrial revolution, they outgrew the human friendly scale necessary to stay imbedded in the village.

The dust, noise, odors and other things associated with large-scale anything required that these businesses locate away from the village. Any time an economic sector moves away from public view, it begins to create its own value system. Any economic sector that's out of sight long enough will begin taking environmental, social, and economic short cuts. The only way to insure integrity in anything is to maintain transparency, where lots of people can see what goes in the front door and what comes out the back door.

The reason corporations were able to pollute with such brashness was because they were isolated from the village. They were big enough to hire security details to keep people away. Snoopers were summarily prosecuted. The old saying "Power corrupts and absolute powers corrupts absolutely" is true for everything but divinity. And I'll lay money that neither you nor I is divine.

After people visit our farm to see our pastured poultry, I always encourage them to try to get a tour at a poultry farm or poultry processing plant. It won't happen. Oh, those industrial folks will hide behind the "biosecurity" excuse, but they really don't want anybody to see it. Like Michael Pollan said in the blockbuster New York Times bestseller *OMNIVORE'S DILEMMA*, if all CAFOs had glass walls, it would fundamentally change the way Americans eat.

The only way to restore integrity in the food system is to re-imbed the butcher, the baker, and the candlestick maker in the village. That means that more farming, industry and processing must revert to a human scale. I am not a hater of big. But I refuse to buy the line that scale is amoral. The same thing done small cannot just be done on a huge scale. Lots of things are that way. Your kitchen is built to produce perhaps 20 meals a day. Just imagine producing 100, then 1,000, then 10,000. You'd have beans slopped on the ceiling and potato skins on the floor. Scale is a factor in anything.

If scale isn't important, why do private and magnet schools advertise small classes? When we say "intimate setting" we mean away from crowds, and private. The same ambiance simply cannot be created in a crowd of 1,000 people. Scale does matter. A 500 acre field of tomatoes is completely different from a backyard garden bed of six plants. The ecology is different. The labor requirement to plant, tend, and harvest is different. That 500 acre tomato field really has nothing in common with that backyard bed. A backyard bed of radishes is far more similar to the bed of tomatoes than the 500 acre field. If a business can become too big to fail, what about all the little businesses that have to suck it up or go bankrupt because they don't merit a place at the taxpayers' trough? Scale obviously changes things.

As Polyface has grown, I've become increasingly concerned about adulterating our core business values. It bothers me that most successful small businesses eventually sell to some big conglomerate. The more I've thought about this, the more I realize that my business values are completely incompatible with Wall Street. As a pre-emptive strike against being swallowed up in the future by some empire, we created a 10-point value definition to keep us true.

1. No Sales Targets. Setting sales or marketing targets makes a business look at its employees differently, its products differently, and its customers differently. It's kind of like a church that sets membership goals: the message is no longer as important as getting sign-ups. What we're willing to compromise to "make the sale" is much greater when a sales target beckons.

Due to quota expectations, sales departments do everything from lying to customers to fudging on sales reports. Sales should be an organic outgrowth of product and service quality. If product or service are good enough, sales will occur. Notice I'm not saying to dispense with marketing. You need to let the world know you exist. But the moment you create a sales target, those notches in your hatchet become the benchmark of success rather than success being a natural outgrowth of your product and service quality.

Setting goals with soul may sound counterintuitive, but it follows what we all know deep down inside: the best things in life are free. Would anybody argue that financial success is better than a happy marriage? We all intuitively understand that salamanders with four legs are better than ones with three, and yet chemical companies selling pesticides or herbicides measure success only in terms of sales volume. Their accountants don't ask for salamander legs.

Here's the question: "What goals are noble enough to justify my life?" That leads to noble and sacred goals, like healing the land, healing employees, healing customers. Goals need to be far bigger than sales. If we strive to be good above all else, growth tends to take care of itself. Growth can also occur in many ways besides gross sales or net profits. We can grow in relationships, knowledge, quality of life, spiritually.

2. No Trademarks or Patents. This idea comes directly from community building and transparency. Such a lunatic notion certainly makes most people cringe. And yet the world survived for a long time without patents. This all goes to open sourcing. What would a business climate without patents look like?

Probably much smaller businesses with much more innovation and less protectionism. Rather than litigating to protect the past, we'd all be scrambling to create the future. But what about competition or copycats? I figure that if I can't stay ahead of the copiers, then I don't deserve to stay ahead. If you study innovation, the ones who are out in front have already gone through a learning curve.

While copiers can shorten the curve or change its trajectory, they still have to go through it. This attitude keeps me lean and learning rather than bureaucratic and superficial. Imagine if everyone had to depend on their own cleverness to stay ahead of the competition. Talk about innovation.

At Polyface, we share everything. We give our rations away. We give our infrastructure designs away. No secrets; it's all open source. Is that foolish? By some counts, thousands of farms now copy what we do. Are we scared? No, because every business that copies our model will heal another few acres. We're much more concerned about healing than competition. A business devoted to healing tends to preserve its patron base. And what a great story.

Yes, we've had numerous people misuse or abuse our concepts. I've visited plenty of farms touting themselves as using the Polyface method and cringed at the adulteration or out-and-out fraud. But what goes around comes around. And the more I get our information out there, like in this book, the more people will understand what true blue really is. When a person begins taking credit for someone else's achievements, the market will eventually reward the innovator—unless the innovator becomes a graspy, paranoid close-to-the-chest protectionist who wants to decapitate the competiton.

Bottom line: the vulnerability that this notion creates also offers a magnanimous spirit that viscerally demonstrates Stephen Covey's plenty vs. scarcity habit in his *7 HABITS OF HIGHLY EFFECTIVE PEOPLE*. Most people who say they believe in openness actually spend a lot of time protecting their stash.

3. Clearly Defined Market Boundary. No difference really exists between an existing empire and an aspiring empire. A one-salary sole proprietorship that aspires to be an empire will have the same attitude as the business that already has an empire. The bigger the outfit becomes, the less innovative it is, partly because it's harder to turn an aircraft carrier than a speedboat.

I confess to being leery of empires because I haven't seen one yet that seemed fair and honest. Empires tend to bully and abuse, in my opinion. When does a business morph from integrity to scandalous? In my opinion, the day it decides to become an empire. If size never registers on your company radar screen, you can become pretty big without selling your soul. But the day you aspire to be the biggest player is the day you begin disrespecting the other players. How about aspiring to be the player that practices hardest? That gives other aspiring players the best hand up to join you in the winners' circle?

At Polyface, we define our market area as within four hours. That's as far as someone can come, personally check us out, and return home in a comfortable day. One of the greatest confirmations that this parameter can be great for business is that this is why Michael Pollan ended up visiting our farm—I refused to send him a T-bone steak. That conviction so piqued his interest that he came, saw, and wrote.

Never underestimate the good things that can happen when you establish a business conviction and then stick with it. Believe it or not, people still appreciate outfits that believe in things.

The emotional freedom that this boundary affords is palpable. When someone calls from Indianapolis or Boston, I'm not even tempted to service them. If I were not a lunatic farmer, I'd be turning myself inside out trying to figure out how to get a pallet of pork bellies to someone in Boston. Instead, I just smile and say, "Find your local land healing farmers and patronize them." It's like Staples: "That was easy."

Dad always said to be leery of people born with a big auger. When someone comes to me with a business idea and starts talking about franchising or being in every major city before he even has a prototype, I'm not impressed. I assume the business will either crash and burn early, or it will be successful but adulterated.

Why would anyone aspire to have an empire? The people who run them seem fairly unhappy. I realize this can't work for every business, but I admit to being disgusted when I hear corporate cheerleaders talking about dominating the world market or beating their chests because they have the largest market share. How about making room for others?

4. Incentivized Work Force. We do everything possible to not have employees. I don't mean we're against help, or against teams. But I'm a fan of bonuses and commissions. I don't even believe in children's allowances—nobody should get paid for breathing.

I like sales commissions. I like bonuses for bringing new customers. Instead of buying advertising, we give $10 credits to any patron in our buying clubs who brings us a new customer. Some patrons bring a couple per drop and end up spending less than if they bought their food at the industrial supermarket. If a patron brings us 10 customers in a year, each of whom spends $1,000, that $100 in free food yielded a $10,000 return. I don't know any advertising agency yielding better returns than that. That's real narrow-cast advertising.

Polyface now rents several farms, and we're placing former interns and apprentices on those farms as independent contract growers. Some grow only for us since they don't want to market. Others grow some for us and some to market themselves. Others grow some for us and then develop an entirely new product that either we can market or we market collaboratively.

Most farms like ours, experiencing growth, would just hire minimum wage employees to do all the work. But we want the best and brightest. The way to get them is to provide open-ended income potential and lots of personal autonomy. Each of these young farmers runs a business in their own rite. Compare that to poultry companies that won't even let a farmer have a backyard flock of chickens for his family's personal consumption. Paying by the piece up to an agreed-on number enables these young people to start farming without any capital. I wish we could also offer them open-ended production capacity, but we must stay within our market volume.

Each agreement is customized based on the young person's goals and desires. And certainly sometimes the landlord's preferences enter into what's doable. All of these growers have been through the Polyface apprentice or intern program, so they've been vetted by working alongside us over an extended period. Now that our farm has developed these prototypes, the next challenge and innovation opportunity is to duplicate and scale them in an anti-Wall Street approach that preserves all the social, ecological, and economic integrity. That will take every bit as much creativity as developing the prototypes in the first place. This incentivized model truly rewards achievement while also delegating responsibility.

Can you imagine what would happen in America's public schools if graduates who had been out for, say, ten years could rate their teachers and the top ten percent received a $50,000 bonus check? And the bottom ten percent were fired? Talk about accountability. When I present this idea to our local school board, the public teachers accuse me of hating education. All they hear is "fired." They don't hear the incentivized bonus. I don't go to these meetings much because it's too painful to see the protectionism surrounding deadbeats.

Rather than spending a bunch of money on trendy advertising and catchy public relations firms, why not redesign job descriptions to create such an enthusiastic, empowered workforce that those ads and PR outfits are unnecessary? We live in a culture that loves minimalism and just-get-by-ism. I think too often we create that spirit by being too timid to make innovatations in our compensation packages that give eager beavers more than a pat on the back.

I have no wage or salary aspirations. I'm happy if key players in our business earn more than I do. That's fine with me. What would I do with more money anyway? I'd just get stuffy, materialistic, and fuzzy-headed. I'd rather go to the grave a pauper but loved by my people, than go wealthy and unloved. Perhaps if more CEOs were less materialistic, their workforce would also show more noble values. Maybe if CEOs were less concerned about their compensation package, employees would be less likely to pinch paper clips from the company stash.

5. No Initial Public Offerings (IPOs). I can hear breath sucking in again. We're going from lunacy to insanity here, aren't we? While this may sound like sacrilege and we all know how growing businesses are starved for cash, consider how many have lost the edge of their good qualities after suddenly becoming flush with cash. I honestly don't know how all of these value ideas work in a capitalistic society, but I know the danger of huge cash infusions.

At Polyface, we've been starved for cash more years than not. And yet that is exactly what makes us innovative—we're hungry. And when we're hungry, we're much more creative. Think about Rikki Tikki Tavi in Rudyard Kipling's famous tale. Hunger made him successful agains the cobra. When we need some capital, we appeal to our patrons to give us short-term no interest loans and they love to invest in something noble.

If your product or service is good enough and your mission noble enough and your cause life-changing enough, you can find other ways to raise capital besides an IPO. This slower, more relationally oriented, pay-as-you-go growth is inherently more organic. Growing from the inside out rather than the outside in follows a more natural pattern. Plants and animals can't grow beyond their ecological resource base, their support structure. When we violate that principle in nature, we get lots of growth and no quality. Bushels of junk. I think IPOs enable a business to grow beyond its support structure, which includes its market, its knowledge base, and its relationship network.

Being satisfied with organic growth keeps us real. If this one principle were used across America's business landscape, we would probably have fewer corporate scandals. Meteoric rises usually result in meteoric falls, so beware fast cash and the imbalance it usually creates. Back to the born with a big auger mentality. An auger is commonly used to bore holes, like oil wells. A guy with a big auger is never satisfied and will often bore through sensitive things to get what he wants.

6. No Advertising. Amazingly, even the largest companies in the world still receive more than 50 percent of their business by word-of-mouth recommendation. That's quite astounding when you think about $2 million for 30-second Super Bowl ads.

At Polyface, we turn existing patrons into evangelists by rewarding them with free product and hugs and slobbery kisses when they bring us new customers. Our culture is starved for appreciation. Showering our good customers with gratitude is the most efficient way to market.

Word of mouth may not be flash-in-the-pan patron building, but it always gets the best quality customer. Only good customers are good for business; bad customers are a drain. If new customers aren't coming, I don't assume advertising is the answer. I assume my products or services aren't where they ought to be; they aren't compelling another patron to join us. I assume if we're in a slump, our marketing problem is probably poor quality and service, not an anemic advertising budget.

By now, some of you may be either livid or taking on air—especially if you run an ad agency. Not to worry. Only 1-3 percent of businesses are on the lunatic fringe anyway. Most will never adopt this. You're safe.

7. Stay Within the Ecological Carrying Capacity. Numerous people have encouraged Polyface to become the Tyson of pastured poultry. But that does not take into account ecological carrying capacity. Manure, guts, blood, feathers and whatever should be metabolized on site. This forces us to decentralize, stay divested across the landscape, and remain aesthetically and aromatically attractive.

We won't join the global trafficking of body parts and offal. In California, where confinement dairy manure is such a problem that they are burning it to power utility plants, such an assault on local ecological carrying capaicty exceeds description. The imported cows are fed imported feed to make manure to be exported to utilities exporting power. The fields that grew the feed are deprived of the manure that rightfully belongs to them in nature's recycling economy. In this industrial paradigm, nature's smoothly operating recycling program creates waste streams and disease.

The local ecology includes the people resource. A business that can't or won't hire people from its neighborhood is not operating within the carrying capacity. Redesigning the business to fit, to nest into the local ecology takes innovation, but anything less will create social, environmental, and pathogenic upheavals. Appreciating our landscape and people resource ecology and staying within those parameters is simply the foundation to being a good citizen and good neighbor.

8. People Answer the Phone. This may seem nitpicky, but how many of us love talking to a robot? After being on the phone with a robot and starting over the fifth time, I can't help but wonder if it wouldn't be more efficient to just hire an incentivized person to answer the phone and deal efficiently with the transaction. I'm convinced that if businesses put their money in people instead of the latest techno-gadget, they would have a lot more happy customers. To be clear, I'm not talking about voice mail; I'm talking about robots.

I answered the phone the other day and a lady on the other end stuttered: "Oh, is this a real person? I didn't expect to talk to a real person." If I were king for a day, I would outlaw these robotic answering devices. Does anyone enjoy hearing "Press 1, press 2, press 3?" We all hate it, and yet businesses succumb to some sales pitch and the assumed efficiency and buy into anti-human treatment. The last time I checked, though, most of us are in business to deal with humans.

A well trained, pleasant-voiced, empowered person can handle my frequent flyer miles redemption ten times faster than the robot. If the airlines would offer a human transaction instead of a robot, I'd gladly give an extra 5,000 miles per transaction for the efficiency and warm fuzzy. At Polyface, we're committed to never having a robot answer the phone.

9. Respect the Pigness of the Pig. We talked about this several chapters ago, but this is one of our core business values that protects us from becoming an empire. Any growth that occurs must harmonize with preserving the pigness of the pig. That includes appropriate seasonal changes, including both production and shelter style.

In the winter, the laying hens come into hoophouses because snow and chickens don't mix well. But once the snow is gone, they go back out onto pasture. We don't raise broilers at all in the winter. Historically, spring was the only time enough extra eggs were laid to have enough for hatching. During late summer, fall, and winter, the household needed every egg just to keep up with the kitchen needs. This appreciation for the ebb and flow of seasons and cycles feeds our emotions with down times and spring times.

Industrial animal operations, in contrast, run full bore all the time. No breaks. Consequently, workers burn out. Owners burn out. Children don't want any part of it. On our farm, in the winter we spend days just lounging around the fire reading books and playing board games. Yes, we sprint in the spring, summer, and fall, but we always have that winter light at the end of the tunnel.

We have many customers, especially chefs, who push us to defy the seasons, build a confinement poultry house, and go into year-round production. But that would not only compromise our pastured poultry integrity, it would put us on a treadmill. I recently visited a large e-corporation and all the employees I talked with were frustrated that they could never get breathing room. The pace was faster each month; expectations higher.

If a business enjoys cyclical movement, it will energize everyone's batteries. The assumption that scaling up the corporate ladder requires us to sacrifice our families and marriages is an unrighteous, evil axiom in America. Our frenetic, work-aholic lifestyles, contrary to popular opinion, are highly abnormal in the continuum of human history. The times of our lives will always trump the paychecks of our lives.

10. Quality Must Always Go Up. Finally, as we grow, we must never compromise quality. Plenty of great small businesses grow up to be ho-hum big businesses. Whatever growth occurs, it can never happen at the expense of quality.

On our farm, one of our primary goals is that every year, we must have more happily dancing earthworms. Kind of the ultimate agronomic shindig. If the earthworms are happy, everything else falls into place. That goal drives how we handle manure, where we put animals, how we handle the landscape. It drives everything. It's all about faithfulness.

As we grow, our suppliers should be happier. Our team members should be happier and more enthusiastic. Our customers should be more loyal. Our water should be purer. Our service should be better in every way. And our products should last longer, cause less pollution, stay out of landfills easier. At the end of the day, does any facet of our business require us to do some fancy talking? Maybe pull up a partition to hide something. Maybe keep us from full disclosure. Require cleverspeak?

I'm reminded of Tyson's claiming "Raised without antibiotics" on chickens when they figured out how to inject antibiotics into the chick embryo before it hatched. Integrity does not require wordsmithing and cleverspeak. It just is or isn't. I want our quality to keep moving up, creating more aha! moments at the dinner plate.

These values are certainly not consistent with Wall Street values. But they will keep a business localized. They will keep the butcher, baker and candlestick maker at a village-friendly scale. And they certainly are opposite the global agenda.

Just to drive the local potential home, remember that America has 35 million acres of lawn. That's land that routinely gets fertilized, irrigated, and then mowed with heavy metal operated by petroleum. If all that ground were turned into edible landscaping, it could feed the nation. When I speak on college campuses, my favorite challenge is for institutions to convert the landscape from inedible to edible. Students could graze on pears, apples and strawberries as they strolled from class to class.

This grows the economy from the inside. The industrial food system extracts wealth from the countryside and translocates it to urban centers. When the food is grown, processed, marketed, distributed, accounted, and eaten locally, the dollars turn over and over, driving appropriate development. Big box stores are net economic losers for a local community, not net economic gainers. Corporate headquarters, with inflated CEO salaries, siphon off the lion's share of what's left over after the Chinese are paid for making the stuff.

Think about this: if a community moved its less than 5 percent in-state food share to just 50 percent, imagine what that would do to the vibrancy, integrity, and plain old security of the food system. Close your eyes and take an imaginary walk down the aisles of the local supermarket. What could be grown there within 100 miles? Just start naming off things. You'll find that 90 percent of what we eat can be grown locally. That would require us to eat seasonally, and to preserve in season for out-of-season consumption. Why is that so absurd? It's normal and reasonable. People have been eating that way for way longer than they've been eating without regard to seasons or local climate.

Our farm is a rural revitalization engine. Money flows into the rural area, not away from it. That's money that stays here in local banks to lend out to local businesses and people who want to buy houses. We're the lion's share of two other local business, the abattoir that we now co-own, and the little Amish feed mill I spoke of earlier. That local economic network recycles these dollars many times. If nobody would take them and spend them at the supercenter, they'd stay here even longer.

Exports are not evil. Imports are not evil. But an economy or a culture dependent on them has neglected the foundation of a strong domestic economy. A nation that can't feed itself is vulnerable. So is a state. So is a community. And so is a household. A well stocked pantry, I would argue, is a matter of national security. The ability of a locality to withstand shocks, whether they be economic, environmental, or societal, is the measure of its strength.

The nonchalant attitude that most people exhibit toward food is appalling. The average community now only has three days' food on hand. Three days! In our house, we could go for at least a year. Granted, we'd be down to some fairly slim pickin's. But if it's the same stuff or shoe leather, I'll take the same stuff. If our food system were localized as much as possible, we wouldn't need all the oil to transport all the stuff. And if we didn't need all the oil, maybe we wouldn't have to fight for it. And if we didn't have to fight for it, maybe we wouldn't have to spend so much money on the military industrial complex. And if we didn't have to spend all that money and put all those soldiers in the field, maybe we could turn our swords into plowshares and be happier.

That kind of talk, of course, is ridiculous, according to most. But knowing that our farm is moving the culture in that direction, even if it never happens or happens way in the future, is still part of the sheer ecstasy of being a lunatic farmer.

TAKEAWAY POINTS

1. An empire is really an attitude.

2. Localized food creates security.

3. The U.S. thrived for 150 years without an IRS.

SUMMARY

L unatic farming: it's a wonderful life. Every day I feel blessed to have known about and then embraced this path.
I could have grown up in a home that ridiculed unconventional thinking. I could have been born in a city. I could have believed the school guidance counselors who said farming was for idiots. I could have majored in institutional agriculture and graduated with peer dependency. I could have become disparaged when friends and co-workers thought leaving steady outside employment to return to the farm was the most foolish idea imaginable. I could have married a ditz. I could have been in a family that made it hard for the next generation to own a piece of the action. I could have alienated my family. I could have been scared to take risks.

This list could go on and on, but you get the picture. Every day I'm deeply grateful for pursuing this lunatic path. As a result, I don't share the despair of virtually every conventional farmer I've ever met. They are fearful of people visiting their farms. After all, people bring disease. They are fearful of prices plunging. They are fearful of peak oil and being unable to acquire chemical fertilizers, pesticides, and herbicides. They are fearful of lending rates. They are fearful that their children will dislike the farm and that they will grow old with no laughter in the house.

Today's conventional farmer lives in a world of fear. Indeed, perhaps we could say our entire culture lives in fear. In sharp contrast, I feel like I live in forgiveness. I don't wake up every morning worried about an epizootic, economic disaster, or ecological calamity. Instead, I can't wait to go out and make animals happy. To participate in a wondrous ecological dance. To embrace my ecological umbilical, and to appreciate that things are right in my world because I have endeavored to create forgiveness and resiliency.

The soil will become more fertile today because of the compost and mob grazing I orchestrate. The animals will be healthier because of the new salad bar I'll give them. Water will flow abundantly because I've invested in ponds to hold back runoff that would otherwise erode soils and harm neighbors downstream. The food I produce will exhilarate three trillion bacteria in every single one of my patrons.

Our farm's economic viability will increase today because we've spent less than we took in, paid our bills on time, developed multiple use infrastructure, and leveraged biomimicry. My spirit soars as I step out into the fresh-scented morning air. The cows lounge contentedly in their paddock, belching, chewing cud, fermenting their salad bar, and sequestering carbon. Earthworm castings fresh from the night's deposits remineralize and purify the soil. Thousands of chickens wait expectantly for their fresh salad bar and insect buffet.

Stacking, synergy, symbiosis: this is the stuff of ecological bounty and forgiveness. The weed-free pastures, vigorous forage, slick cattle, and smiling workers testify that this system is indeed sheer ecstasy. I do not anticipate an oops day. Industrial/chemical farming has had several oops days.

Remember when confining animals in muddy lots brought on hog cholera and Newcastle's disease? Or feeding distiller's grains to dairy cows brought on undulant feaver and brucellosis? How about when everybody realized DDT and other chemicals were responsible for three-legged salamanders and infertile frogs?

How about when feeding dead cows to cows was the smoking gun in bovine spongiform encephalopathy? And how about when Cdiff and MRSA, commonly known as hospital superbugs, were linked to antibiotic feeding on factory farms? Remember when subsidized corn turned into cheap high fructose corn syrup was fingered as the primary cause for Type II diabetes? And just around the corner, devastating ecological and nutritional results from genetic engineering.

I'm thankful beyond words-ecstatic, in fact-that I won't have any of these oops days. Just so nobody misses the point, what caused these oops days? Wrong headed thinking, that's what. And it was fully endorsed, promoted, and financed by credentialed scientists, government programs, and pride. Every single time, those of us who refused to go along, who refused to sign up, and who questioned were summarily rejected as lunatics. Well, I think we need more lunatics. If that's what it takes to espouse the truth, then I want to be one.

I hope as we've examined the clash of paradigms you've embraced, with appreciation and confidence, the lunatic approach. I don't know what it will take for the current lunatic approach to become mainstream. I think every culture and every historical period has had its lunatics. They laughed at Fulton. Galileo was thought to be possessed. Columbus was going to fall off the edge of the world. Native Americans were barbarians. Slaves were not humans. Human embryos are not alive. Home schoolers are social miscreants. America had to liberate Iraq.

I'd better stop before I make everyone angry. The point is that when you examine history, more often than not the majority view is incorrect. In fact, when people ask me about my position on pieces of legislation, I've come up with an efficient way to make a decision. Any more, laws are too long to read anyway, so I just look at who supports them. If Monsanto is for it, I'm against it. Just look up what Monsanto supports and the opposite will be the lunatic fringe-and the right position. It's really that simple.

A lot of people ask me what I see in the future. I don't prophecy. One of my hobbies is collecting the prophecies of experts that turned out to be ridiculously wrong. Experts-especially popularly endorsed experts-are wrong more often than you can imagine. So I have no idea where things will be in five years. Are we lunatics going to be completely muzzled and shut down by the current conventional paradigm? I have no idea.

But I do know that long after the ecological adulterers and prostitutes have run their course, earthworms will still want organic matter to eat and will still create fertile soil out of poverty. I know cows will still want to eat forage. Chickens will still want to chase bugs. Pigs will still want to dig for acorns. And the three trillion member community inside each of us will still want to eat things you can pronounce, that you can make in your kitchen, and that will rot. Compost will still grow the most nutrient dense vegetables.

I do not worship at the altar of science when science despises these natural laws. I do not believe for a minute that genetic engineering will save humankind or that mono-cropping can ever be made more productive per acre than diversified synergistic symbiotic relational farming. I do not believe animal factories can ever be more efficient, productive, or healthy than pasture-based and deep bedded models. Such prejudice, of course, puts me firmly in the anti-science lunatic camp.

Ah, what a wonderful place to be. Resting in the principles that have proven themselves for millennia. Resting in the historical authenticity of food communities throughout the world. Beautifying landscapes. Keeping customers out of the hospital. Yes, that's where I want to be.

Welcome home to the sheer ecstasy of being a lunatic farmer.

Notes:

Index